Upper Carboniferous Fossil Flora of Nova Scotia

Upper Carboniferous Fossil Flora of Nova Scotia

In the Collections of the Nova Scotia Museum; with Special Reference to the Sydney Coalfield

Erwin L. Zodrow and K. McCandlish
College of Cape Breton

Nova Scotia Museum
Halifax, N.S.
1980

ISBN 0-919680-08-9

Printed in Canada

Published by
The Nova Scotia Museum

as a part of
The Education Resource Services Program
of the
DEPARTMENT OF EDUCATION
Province of Nova Scotia

Hon. Terence R. B. Donahoe
Minister

Gerald J. McCarthy
Deputy Minister

Halifax, Nova Scotia
1980

Produced by the
Nova Scotia Communications and
Information Centre, Box 2206
Halifax, N.S. B3J 3C4

Contents

For Friedrich von Glasow

DR. CHESTER A. ARNOLD

1901-1977

Professor Emeritus of Geology and Botany,
and Curator Emeritus of Paleobotany,
the University of Michigan
Paleontological Museum,
Ann Arbor, Michigan,
United States of America.

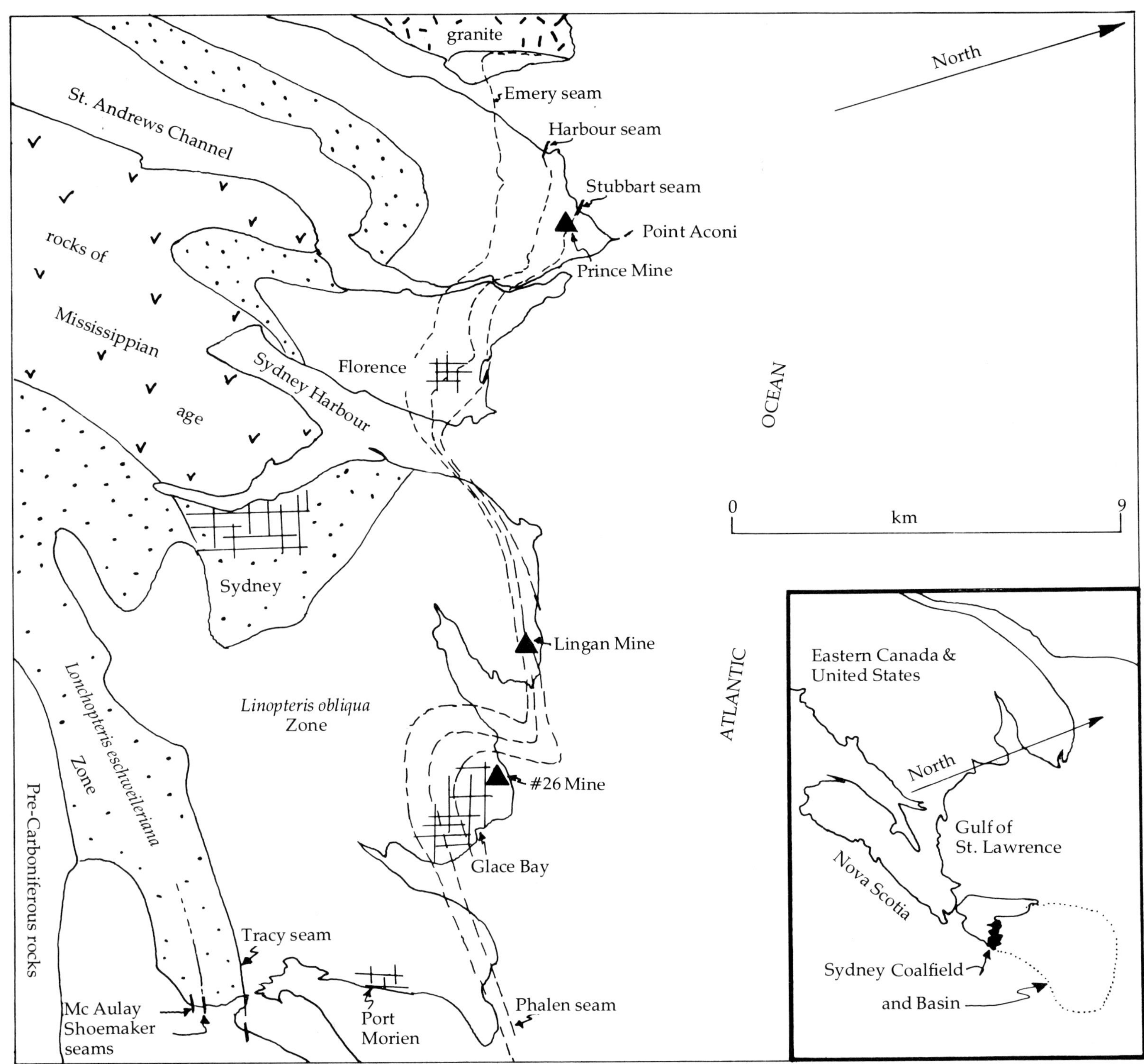

Sydney Coalfield, Nova Scotia, Canada, in relation to eastern North America.

Background

The Nova Scotia Museum geological collections originated in 1831 when the Halifax Mechanics Institute came into being. However it was not until 1868 that legislation was passed which created the Provincial Museum. Collections of the Mechanics Institute were combined with those on display at the exhibitions in London 1862, Dublin 1865 and Paris 1867.

The Carboniferous flora specimens from these collections remained in cabinets for well over 100 years with limited identification and only a handful of specimens identified to the species level.

Dr. Erwin Zodrow of the College of Cape Breton had collected a large number of fossils from the Sydney Coalfield by late 1975 and was anxious to have taxonomic work done on these specimens. A proposal was made to this museum by Dr. Zodrow to invite Dr. Chester A. Arnold, Professor Emeritus in Botany at the University of Michigan Palaeontological Museum at Ann Arbor, Michigan, to identify to species level the specimens collected by Dr. Zodrow and the old specimens collected in the 1800's and housed in the museum at Halifax. In return Dr. Zodrow is acting as collector for the Nova Scotia Museum in the sense that Carboniferous fossils collected by him from the Sydney Coalfield are the property of the Nova Scotia Museum but are on permanent loan to the College of Cape Breton under the direction of Dr. Zodrow. All specimens on permanent loan are identified by addition of 977G to the F-number and future specimens will be identified by 978G, 979G, etc. Included under the 977G numbers are samples of coal-associated minerals listed in Appendix I and coal samples presently housed in the glass display table at the College of Cape Breton.

I wish to express my gratitude to Dr. Erwin Zodrow and Mr. Keith McCandlish for presenting the opportunity for this museum to have such an excellent collection, and to Dr. Chester A. Arnold who applied his expertise to our specimens.

R. G. Grantham
Curator of Geology
Nova Scotia Museum
Halifax, N.S.
1979

Foreword

Wer die grossen Schwierigkeiten kennt, welche mit der Bestimmung der Versteinerungen, und ueberhaupt mit der Bearbeitung dieses Gengenstandes verbunden sind, der wird sich geneigt fuehlen, auch den gegenwaertigen Versuch zur Erweiterung unserer Kenntnisse in Ansehen der Versteinerungen mit der naemlichen Schonung und Nachsicht aufzunehmen, welche die frueheren Beitraege bereits erfahren haben.

(He who knows of the great difficulties connected with the definition of petrifactions and, generally, with the treatment of this subject, will feel inclined also to view the present attempt to enlarge our knowledge of petrifactions with the selfsame consideration and indulgence which were already shown towards earlier contributions.)

— E. F. von Schlotheim
(1820, p. XXXI)

Introduction

Because of its historical aspect there are problems in paleobotany which are well summarized by Beck (1970, p. 175):

"Because plant fossils are usually only fragments, generic names are rarely applied, initially, to whole plants. Furthermore, I know of *no* instance in which a generic name has been applied to a population of whole fossil plants. Because a generic name is likely to be applied to a single piece of preserved tissue, such as a small axis segment or a leaf, or a reproductive structure such as a cluster of sporangia or a seed, or even a single spore, a paleobotanical genus is often little more than a name. Its biological significance increases with discovery of additional similar specimens, and becomes even greater with determination that several organ genera represent the same plant. This conservative, albeit frequently necessary, practice of assigning names to fossil plant fragments results in an unfortunate multiplication of names which may lead to confusion and misinterpretation on the part of paleobotanists and nonpaleobotanists alike."[1]

We gather from this quote that (1) it is difficult to define related plant genera in the context of paleobotany (definition of genera is a function of discontinuity in families and their variations, and a genus connects related species); (2) nomenclature of extinct botanical species is confusing: see EXPLANATIONS FOR ENUMERATION OF SPECIES, p. 20. Adding to the overall confusion are problems concerning the question of classification of true versus seed ferns in the absence of diagnostic features; hence it may be advantageous to create a category which encompasses such a situation (see Table 1).

A different case in point is presented by the question of species differentiation versus conspecificity: e.g., *Linopteris obliqua* vis-à-vis *Linopteris obliqua* var. *bunburii* (= *L. bunburii*). The former is postulated a guide fossil, the latter a species of prolonged duration in the Morien series of the Upper Carboniferous period (Bell, 1938, p. 64). It is believed that both species originated from different positions on a frond, and that they are conspecific, thus the variety becomes a synonym.

In order to create categories for certain fern-like fossil plant specimens that would otherwise be unclassifiable because they are sterile (lack sporangia or seeds, i.e., reproductive organs), form-genera[2] are used as a basis for classification. Form-genera, defined by shapes and venation patterns of pinnules, their relation to fronds, mode of attachment of pinnules to a rachis (pinnule morphology), among others, cannot imply any relation between or among groups of *extinct* fern-like plants. Form-genera, with morphologic characteristics as criteria for classification (Table 1), constitute an expedient and workable scheme, analogous to "working hypotheses" in experimental sciences, by which to classify sterile fern-like foliage. This mode of classification, although rather general, is not based on information relating to natural affinities because of the absence of reproductive structures, and inferences about living ferns are not generally possible.[3] However, the use of form-genera is convenient at times, especially when we speak of "spores borne on pecopterid fronds"; *by definition,* "pecopterid" implies foliar

1 Gothan and Weyland (1964, p. 173-180) approximate the idea of one generic name for one population by using *Calamites* Suckow as "Gesamtgattung": subgeneric taxa:
- a) *Stylocalamites* Weiss
- b) *Eucalamites* Weiss
  - b1) group of *C. cruciatus* Sternberg
  - b2) group of *C. carinatus* Sternberg
  - b3) Genus *Calamophyllites* Grand'Eury
- b3) contains roots, foliage, reproductive structures and other parts of the plant.

2 Private communication with Dr. Arnold, June, 1976.

3 Informative hierarchical classification schemes are based on anatomy, male fructification, fructification-genera, spore forms or pollen kernels (Gothan and Weyland, 1973, p. 232, 312); refer to Table 4.

Table 1

Provisional classification schema for fern-like fronds of Carboniferous age (Arnold, 1976)[4]. This table is incomplete in respect to genera.

Kingdom: Plantae	Subkingdom: Embryophyta
Division:	Tracheophyta
Subdivision:	Pteridophyllophyta (fern-leaved plants)
Class:	Pteridophylleae
Order:	Pteridophyllales
Family:	Alethopterides
	Neuropterides
	Palaeopterides
	Pecopterides
	Sphenopterides

Form-genera under the families

Alethopterides	Neuropterides	Palaeopterides	Pecopterides	Sphenopterides
Alethopteris	*Cyclopteris*	*Adiantites*	*Alloiopteris*	*Diplotmema*
Lonchopteridium	*Linopteris*	*Aneimites*	*Callipteridium*	*Palmatopteris*
Lonchopteris	*(Dictyopteris)*	*Anisopteris*	*Callipteris*	*Rhodea*
Megalopteris	*Neuropteris*	*Archaeopteridium*	*Emplecopteris*	*Sphenopteris*
		Cardiopteris	*Lescuropteris*	
		Eremopteris	*Mariopteris*	
		Palaeopteridium	*Odontopteris*	
		Rhacopteris	*Pecopteris*	
		Sphenopteridium		
		Triphyllopteris		

Remark: "The subdivision Pteridophyllophyta contains fern-like fronds of which many cannot be assigned with absolute certainty either to the true ferns or to the seed ferns".[4] Therefore, in the absence of knowledge concerning reproductive organs, or mode of reproduction, it is difficult to establish natural affinities.

The genus *Dictyopteris* Gutbier is listed for historical perspective only; it is a pre-occupied generic name for brown algae. Presl in 1838 used the name *Linopteris* (Crookall, 1959, Pt. 2, p. 201).

4 Private communications with Dr. Arnold in June 1976 and in May 1977.

characteristics. It is important to appreciate that classification of Pteridophyllophyta, according to form-genera, encompasses both seed and true ferns *in the absence of our knowledge of diagnostic reproductive organs in the collected specimens.* See Table 1.

The subdivision Pteridophyllophyta contains several "families" which may be traced to Brongniart (1822). In this catalogue, dualism of classification is evident in respect to true and seed ferns (it is not so for pteridophylls); families are classified using foliar characteristics, while some species under these families are classified according to fructification-genera (true ferns), and others by male fructifications (seed ferns) (see Tables 2, 3 and 4).

An additional consequence of classifying fossil material according to form-genera is that very often supra-generic categories are uncertain. As Professor Arnold pointed out (*in litt.*, July, 1976):

"Rules that govern the naming of fossil plants include special provisions for these difficult cases by exempting the paleobotanist of the necessity of always assigning a genus to a family, a family to an order, and an order to a class. It is permissible to set up special groups for these form-genera, or, as often happens, a genus is named without reference to any higher category. A good example is *Samaropsis,* a name that can legitimately apply to any fossil seed encircled by a wing. Many *Samaropsis* specimens undoubtedly belong to the Cordaitales, but others to the Pteridospermophyta.

"The problem of fossil plant classification, in addition to the fact that some of the categories used cut across natural groups, is that botanists themselves are far from agreed on what system to use for classification of plants in general. It just means that botanists admit they have not found all the answers."

Since the Linnaean hierarchy (Linneai, 1740) does not carry in itself necessity nor sufficiency (or the logical if-and-only-if relation), an observed sterile fern-like fossil leaf may have originated from either a spore-bearing or seed-bearing plant (a third possibility is provided for by a free-sporing progymnosperm of Beck (1970, p. 176)); hence the arrangement in Table 1. On the other hand

"This matter of perpetuating a purely artificial classification for the fernlike foliage becomes increasingly difficult as time goes on because a considerable number of the pteridophylls can now be included in some natural category, but [it] is not yet feasible to attempt using a natural system for only some of the pteridophylls. . . . So in my opinion it is best to rely entirely on an artificial classification system when using names for stratigraphic citations or in cataloging museum specimens. . . . Also there is a third group that is now in the picture, Dr. Beck's progymnosperms, and only the All Wise One knows how many more are still unrecognized and that someone may recognize within the next hundred years. There aren't many progymnosperms in the Sydney flora, but there are a few that I think may ultimately repose there." Dr. Arnold (*in litt.*, May 10, 1977.)

The genera *Crossotheca* Zeiller and *Zeilleria* Kidston are difficult to accommodate in fructification-families because it is uncertain whether they belong exclusively to true ferns

Table 2

Juxtaposition of some foliar families and fructification-genera of true ferns of the Carboniferous Period.

Division:	Tracheophyta
Subdivision:	Pteropsida
Class:	Filicineae (true ferns, i.e., spore-bearing)
Order:	Filicales
Family:	Sphenopterides (foliar grouping)
Fructification-genera:	
Crossotheca?, Oligocarpia, Renaultia, Zeilleria	
Family:	Pecopterides (foliar grouping)
Fructification-genera:	
Acitheca, Asterotheca, Crossotheca?, Cyathotrachus, Eupecopteris, Ptychocarpus, Senftenbergia, Zeilleria?	
Order:	Marattiales
Family:	Marattiaceae
Genus:	*Psaronius*

Table 3

Juxtaposition of some foliar families and genera, and anatomy- and fructification-genera of seed ferns of the Carboniferous Period.

Division:	Tracheophyta
Subdivision:	Pteropsida
Class:	Gymnospermae (naked seeds)
Order:	Pteriodospermae (seed fern)
Families:	Pecopterides or Sphenopterides*

Genera (by foliar, anatomical and reproductive groupings):

Alethopteris, Crossotheca?, Dicksonites, Eoangiopteris, Lyginopteris, Mariopteris?, Pecopteris, Samaropsis?, Sphenopteris, Zeilleria?

*Refer to Table 4 for different groupings.

Table 4

Juxtapositional schema, according to different botanical criteria, of some families and genera of seed ferns of the Carboniferous Period; modified after Gothan and Weyland (1973, opposite p. 312).

Subdivision: Pteridospermophyta
Class: Gymnospermae (naked seeds)
Order: Pteriodospermae (seed ferns).

FAMILIES (based on anatomy)	Lyginopterideae	Callistophytoceae	?	?	Medulloseae
FAMILIES or GROUPS (based on foliage)	Sphenopterides i.p.	?	Paripteris	Sphenopterides	Imparipteris
GENERA (based on foliage)	*Lyginopteris* · *Sphenopteris* i.p.	*Odontopteris ?* *Reticulopteris*	*Neuropteris* parip. *Linopteris*	*Sphenopteris* i.p.	*Alethopteris* *Lonchopteris ?* *Neuropteris imp.* *Odontopteris ?* *Reticulopteris ?*
GENERA (based on male reproductive structures)	*Telangium* i.p.	*Psaliangium*	*Potoniéa*	*Crossotheca* i.p.	*Whittleseya* *Aulacotheca*

i.p. = in part.

Callistophytoceae bears pecopterid-like foliage.

Mariopteris and *Pecopteris* (in part) must find a place in this table in addition to *Samaropsis* and the male fructification-genus *Zeilleria*, borne in part on pecopterid fronds; see also Table 3.

(Table 2). Gothan and Weyland agreed that both genera may in part belong to the pteridosperms (1973, p. 258). Professor Arnold (MS textbook, 1977, p. 30-31; see ACKNOWLEDGEMENTS, p. 25) believed that *Crossotheca* belongs exclusively to pteridosperms, while the taxonomic status of *Zeilleria* is still uncertain. This uncertainty stemmed from different interpretations of the "oval or spherical spore-bearing organ borne on a short pedicle formed by the extension of a vein beyond the margin of a *Sphenopteris* pinnule" (p. 30). Hence, these two genera are classified as well in Table 3 (seed ferns).

Alternatively, the fructification-genera in Table 2, such as *Acitheca, Asterotheca, Cyathotrachus* and *Ptychocarpus,* for example, can be placed in the family Marattiaceae (a tall tree fern). Similarly, *Oligocarpia* is classifiable under the fructification-family Gleicheniaceae, and *Senftenbergia* under Schizaeaceae (Abbott, 1954; Arnold, 1947, chapt. 8; Gothan and Weyland, 1973, p. 229-257). However, Jennings and Eggert (1972) showed that the classification of the genus *Senftenbergia* under the family Schizaeaceae is no longer tenable if evidence from the Chester Series, Upper Mississippian, is considered. The evidence consists of compressions and pyrite replacement of plant material (from which

Table 5

One possible schema of classification for sphenopsids, lycopsids, and pteropsids of the Carboniferous Period. This table is restricted to collected genera (included are genera from the St. Francis Xavier University Collection, Appendix II).

Division:	Tracheophyta
Subdivision:	Sphenopsida
Class:	Equisetinae
Order:	Calamitales (giant horsetail)
Family:	Calamariaceae
Genera:	
(foliage):	*Annularia, Asterophyllites*
(aborescent sphenopsid):	*Calamites*
(strobilus):	*Macrostachya, Palaeostachya.*
Class:	Sphenophyllinae
Order:	Sphenophyllales
Family:	Sphenophyllaceae
Genus:	*Sphenophyllum.*
Subdivision:	Lycopsida
Class:	Lycopodinae
Order:	Lycopodiales
Family:	Lepidodendraceae
Genera:	
(trunks and branches):	*Asolanus, Lepidodendron, Lepidodendropsis, Lepidophloios*
(leaves):	*Lepidophyllum*
(sporophylls):	*Lepidostrobophyllum*
(strobilus):	*Lepidostrobus*
(rhizophore:)	*Stigmaria*
Family:	Sigillariaceae
Genera:	
(trunks):	*Sigillaria*
(leaves):	*Sigillariophyllum*
(strobilus:)	*Sigillariostrobus*
(rhizophore):	*Stigmaria*

(Table 5, continued)

See *Triletes auritus* for a short statement of macrosporangial affinities.

Subdivision:	Pteropsida
Class:	Gymnospermae
Order:	Cordaitales
Family:	Cordaitaceae
Genera	
(pith cast):	*Artisia*
(male cone):	*Cordaianthus*
(forest tree):	*Cordaites*
(winged seed):	*Samaropsis*

Summary view of subdivisions, classes, orders, and families

Subdivision:	Sphenopsida
Class:	Equisetinae
Order:	Calamitales
Family:	Calamariaceae
Class:	Sphenophyllinae
Order:	Sphenophyllales
Family:	Sphenophyllaceae
Subdivision:	Lycopsida
Class:	Lycopodinae
Order:	Lycopodiales
Families:	Lepidodendraceae
	Sigillariaceae
Subdivision:	Pteropsida
Class:	Filicineae
Order:	Marattiales
Family:	Marattiaceae
Order:	Filicales
Families:	Pecopterides
	Sphenopterides (foliar)

(Table 5, continued)

Class:	Gymnospermae
Order:	Pteridospermae
Families:	Pecopterides
	Sphenopterides (foliar)
Order:	Cordaitales
Family:	Cordaitaceae
Subdivision:	Pteridophyllophyta
Class:	Pteridophylleae
Order:	Pteridophyllales
Families:	Alethopterides
	Neuropterides
	Palaeopterides
	Pecopterides
	Sphenopterides

thin sections were cut) that showed the diagnostic features of *Senftenbergia,* along with sporangial morphology. Information obtained about the vascular system showed in particular that:

a) primary pinnae have a C-shaped vascular system with two embedded protoxylem areas accompanied by parenchymatous peripheral loops (see Gothan and Weyland, 1973, p. 250, Abb. 188(f));

b) secondary pinna axes which bear pecopterid pinnules show a cylindrical and tapered vascular strand with a single embedded protoxylem and peripheral loop.

This anatomy may be compared with terminal parts of sterile fronds of the genus *Ankyropteris* P. Bertrand (a true fern); moreover, sporangia similar to *Senftenbergia* were found on such petrified frond remains. Ankyropterid-type anatomy as in a) and b) above is not closely related to any living family of fern. Hence *Senftenbergia* has affinities with some members of the ankyropterid-type group and is not therefore a schizaeaceous fern.

Refer to Sitar *et al.* (1973) for Carboniferous macroflora of the West Carpathians (genera *Asterotheca, Callipteridium,* and *Cordaites*); to Grauvogel *et al.* (1975) for discussion of fertile fronds from the Stephanian of the Central Massif of France (included are *Asterotheca* and *Senftenbergia*); to Le Roux (1976) for ferns of Lower Permian in the lower Gondwana, and see also Chandra (1974) for India's lower Gondwana; to Archangelsky *et al.* (1973) for Argentina-Brazil fossil flora studies; to Fefilova (1968, 1973) for Permian ferns (included are *Asterotheca, Oligocarpia, Pecopteris* and *Sphenopteris*); to Herbst (1972) for Triassic fern fossils of Argentina, and to Khramova and Pavlov (1971) for Upper Triassic pteridophytes and pteropsida.

While Prof. Arnold took a rather skeptical view in the matter of classifying fern-like foliage, Gothan and Weyland (1964; 1973) tended to be dogmatic in their approach.

Potonié and Gothan in 1921 classified alethopterids and neuropterids as follows, taken here from Gothan and Weyland (1973, p. 10, 303-4):

Spermatophyta
  Gymnospermae
    Pteridospermophyta
      Alethopterides
      Neuropterides.

Under alethopterids, in particular the genus *Alethopteris* as defined by Sternberg, there are the typical (varietal?) forms of *A. serli, A. lonchitica* (see comments under *A. lonchitica* in the catalogue section), *A. decurrens* and *A. grandini;* smaller forms of alethopterids assume a pecopterid-like appearance, as in *A. davreuxi.* In addition, a transition exists between the linear and the meshed veining seen in *Lonchopteris* and *Lonchopteridium.*

The neuropterids are grouped as follows (Gothan and Weyland, 1973, p. 303):

Group 1: Neuropterides imparipinnatae (belong with certainty to Medulloseae).
(a) forking of veins: Genus *Imparipteris* Gothan

(b) meshing of veins: Genus *Reticulopteris* Gothan

Group 2: Neuropterides paripinnatae (this is the group of *N. gigantea;* it may belong to Medulloseae).
(a) forking of veins: Genus *Paripteris* Gothan

(b) meshing of veins: Genus *Linopteris* Presl zum Teil.

Examples of Group 1(a) include *N. heterophylla, N. tenuifolia, N. obliqua, N. schlehani, N. rarinervis* and *N. ovata.* In Group 1(b) a common example is *R. muensteri,* (classified in this catalogue as *Linopteris muensteri*). In Group 2(a) there are only a few species, not of great concern for our purposes, while Group 2(b) contains the familiar *L. obliqua, L. neuropteroides* and others.

Needless to say, Gothan (1973, p. 301, 303) attached great importance to whether parts of fronds (Wedelteile) terminate with one pinnule (Blättchen) (the impar form), or two pinnules (the par form). He stated:

"Diese Neuropteriden waren von Gothan nach dem Wedelaufbau in zwei Gruppen unterteilt worden, die Imparipinnaten, deren Wedelteile mit einem Fiederblättchen und die Paripinnaten, deren Wedelteile mit zwei Blättchen enden und ausserdem Zwischenfiedern besitzen. Es hat sich inzwischen herausgestellt, dass diese Unterteilung nicht nur vollständig berechtigt, sondern auch durch die Art ihrer inzwischen bekanntgewordenen männlichen Fruktifikationen geboten ist. In der Praxis wurde aber die alte Einteilung nach der Aderung weiter benutzt, wobei die fiederig geaderten Formen als *Neuropteris* die maschig geaderten als *Linopteris* Presl bezeichnet werden. Das entspricht aber den jetzigen Kenntnissen nicht mehr. In Wirklichkeit sind die imparipinnaten Neuropteriden mit den Alethopteriden näher verwandt als mit den paripinnaten Neuropteriden. Gothan hat daher die Neuropteriden in vier Formgattungen geteilt bei denen die Namen *Neuropteris* und z. T. *Linopteris* verschwunden sind'' (p. 301, 303).

It is of interest to note that Gothan (1964, p. 257) asserted that almost all species of *Pecopteris* are true ferns (''sind echte Farne''). However, he did not neglect exceptions; if these are reviewed in his textbook one detects distinct Arnoldian skepticism, as when reading Table 3 (keeping in mind Table 1). Gothan recognized the dilemma embodied in Tables 1 to 3 but dealt with fern-like foliage (''farnlaubige Gewächse'') only in terms of pteridosperms (''Nacktsamer'') or true ferns.

Summary

Three principal remarks are in order concerning this collection:

a) the inclusion of possibly new botanical species;

b) known species, and varieties of species, not previously reported from the Sydney Coalfield, and

c) (guide) fossils and their associated flora which explicitly contributed to the establishment of zones of the Morien series and its correlation with European sequences.

A solid triangle (▶) is used in the list of species in the following catalogue to indicate either a) or b) above, or to indicate discovery of a genus not previously reported from the Sydney Coalfield, as compared to Bell's catalogue (Bell, 1962a). The third point merits closer scrutiny.

Available geological maps of the Sydney area issued in 1938 (359A, Bras d'Or; 361A, Sydney; and 362A, Glace Bay) are based on Bell, Hayes, and Goranson's efforts. The subdivision of the sedimentary sequence of assumed Upper Carboniferous age in the Sydney area, containing coal measures either of workable or non-workable nature, is based explicitly on fossil parameters (the presence or absence of a fossil flora). These fossil parameters were defined by Bell in 1938 and employed by him to divide the strata into three floral zones (Table 6).

Bell (1962) stated very clearly why he correlated the L-zone (Lch-zone) with Westphalian C age and the Pt-zone with the Westphalian D. His reasoning is summarized by stating that the boundary between Westphalian C and D, or the contact between the L-and Pt-zones, was established because of the first appearance of a number of species above the Emery seam that are lacking in the L-zone (exclusion principle). Specifically, Bell (1962, p. 5-6) wrote that although *N. ovata* is regarded as one of the best guide fossil for strata of

Table 6

Division of the Morien series in Sydney Coalfield, Nova Scotia, according to the presence of *Ptychocarpus unitus, Linopteris obliqua, Lonchopteris eschweileriana* and associated flora (Bell, 1938).

Surface of angular unconformity	Main Coal Measures	Equivalences of main coal measures across Cape Breton Island: Port Morien, Glace Bay to New Campbellton.
WESTPHALIAN D		
	Point Aconi	*Cranberry Head.*
	Lloyd Cove	*Carr, Paint, Bonar.*
	Hub and Stubbart	*Barachois, Crandel, Chapel Point, Stubbart.*
Ptychocarpus unitus Zone	Harbour	*Blockhouse, David Head, Victoria, Sydney Main.*
Pt-zone.	Bouthillier	*Fairyhouse, Willy Fraser, Franklin, Millpond.*
	Phalen	*Mc Auley, Gowrie, Lingan, Last Chance — Stanley, Toronto, Collins, Six Foot.*
	Stony	*Lower Jubilee.*
(roof of the Emery seam)		
WESTPHALIAN C		
	Emery	*South Head Wilson, Spencer, Emery Ross, D. McGillivray, Henderson.*
Linopteris obliqua Zone	Gardiner	*Long Beach, O'Dell, Young, Ingraham, Gannon Elkins.*
L-zone.	Mullins (Ormond)	*Coalbrook, Le Cras Wagner, Martin, Fitzpatrick, Carroll, Mullins Fraser, Mullins.*
(roof of the Tracy seam)		
WESTPHALIAN C		
	Tracy	
Lonchopteris eschweileriana		*Shoemaker.*
Lch-zone		*Mc Aulay.** (see Fig. 9)
		Round Island.

*May be a spelling error on the geological map, but we maintain this spelling for consistency. (Courtesy Coal Division, Cape Breton Development Corporation, 1954.)

Westphalian D age, the appearance of this species near the top of the L-zone does not necessitate lowering the boundary between Westphalian C and D. In addition, *L. obliqua, N. tenuifolia,* and others were not reported from the Pt-zone by Bell (1938).

With the development and exploitation of new mines in the Sydney area (Prince Mine at Point Aconi mining the Stubbart seam since 1975, and Lingan Mine at Lingan mining the Harbour seam since 1972) new fossil material has been collected. As indexed in this catalogue, abundant specimens of *Linopteris obliqua* were obtained from these underground sources; these coal measures are located in the Pt-zone (Table 6). Moreover, one specimen of *Lonchopteris eschweileriana,* F-390, was collected in the roof of the Emery seam which forms the basal member of the Pt-zone. The inescapable conclusion is that some of these species were contemporaneous in Upper Carboniferous time. By implication we must assign *L. eschweileriana* to the entire Morien series although admittedly this species is

Table 7

Coal measures of Sydney Coalfield in approximate correlation with the Pennsylvanian series of Nova Scotia (Cumberland to Morien), Pennsylvania (Pottsville to Conemaugh), and Europe (Westphalian to Stephanian stages).

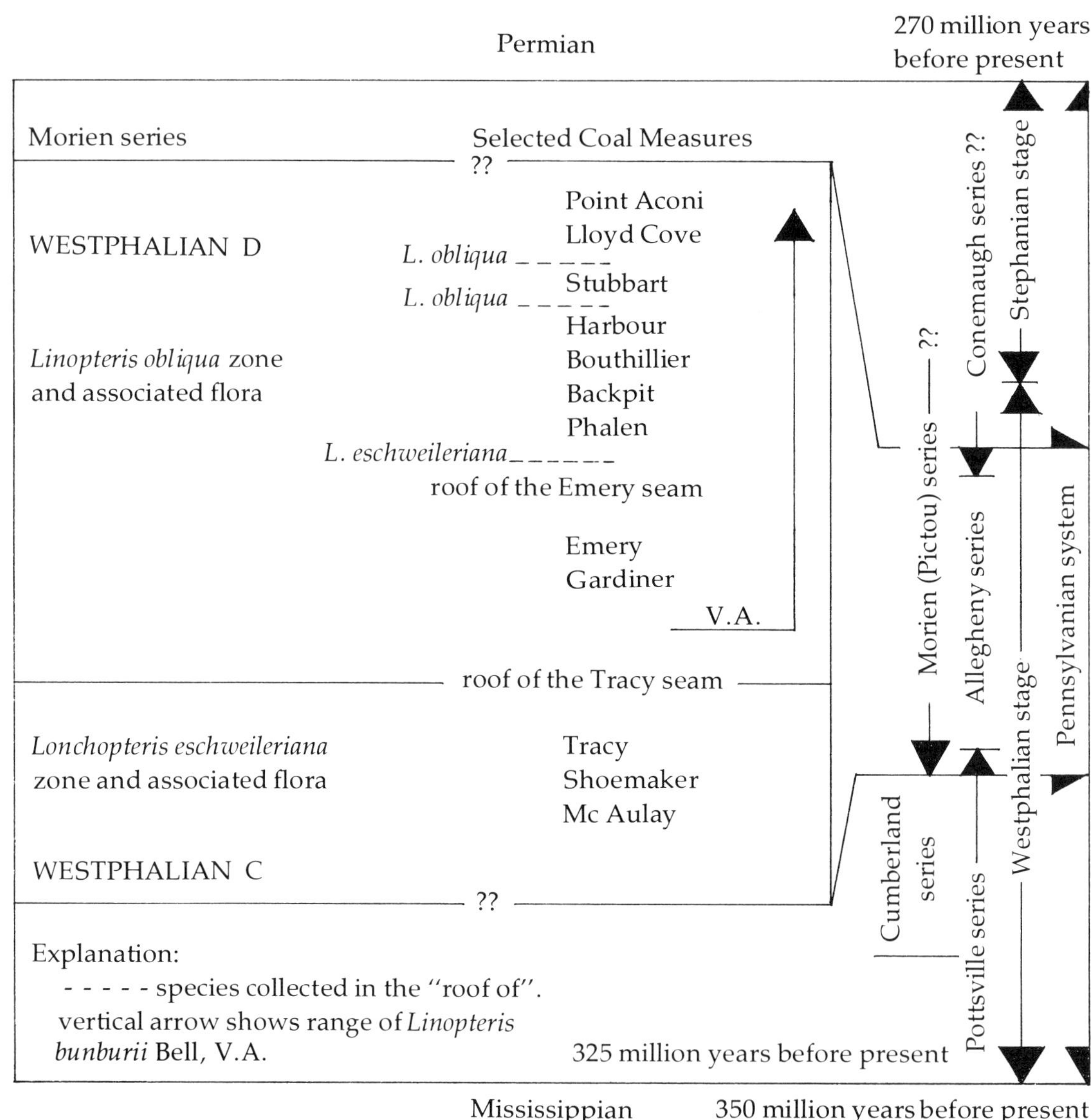

Note: the Namurian stage, at the base of the Westphalian stage, has been omitted. Ages in millions of years are given for the beginning of the units.

extremely rare in the Sydney Coalfield and awaits discovery in the L-zone. Therefore, Bell's use of these fossils (i.e., *Ptychocarpus unitus, Linopteris obliqua,* and *Lonchopteris eschweileriana*) as guide fossils is no longer accurate (Table 7). It follows that these interrelated points must necessarily be considered: a) the boundary line of floral zones between the L- and Pt-zones; b) elimination of one or combination of the L- and Pt-zones, and c) relocation of the Westphalian C-D boundary. The presence of *L. obliqua* in Bell's Pt-zone makes the division line between Bell's L- and Pt-zones obsolete and the floral zones in Table 6 need reinterpretation in the light of the new fossil finds. There do not seem to be compelling reasons to relocate Bell's boundary line between the L- and Pt-zones above the Stubbart seam; rather, the authors favor elimination of the Pt-zone thus proposing two floral zones as shown in Table 7. Bell's contacts which define the floral zones shown in Table 6 may be interpreted as lines that mark the first recorded appearance of characterizing species, since no *P. unitus* has been found in the lower strata, and similarly no *L. obliqua* in the older strata. This is not synonymous with and does not imply first botanical appearance of a species, the basis for Bell's zonal division. In fact, one never is sure that the first record is even close to the first existence of any species.

Bell drew the contact between the L- and Pt-zones and the boundary between Westphalian C and D to coincide; this position is difficult to defend and he had difficulties himself in the evaluation of the flora associated with the *P. unitus* zone (1938, p. 1, 17).

We give the following reasons for the relocation of the Westphalian C-D boundary as shown in Table 7:

a) the occurrence of *L. obliqua* in Pt-zone counters Bell's argument of 1938 (p. 12-16), i.e., that the absence of this species from the Radstockian (Westphalian D) in the British coalfields was one important factor for his assigning the L-zone to Westphalian C;

b) key fossils associated with *L. obliqua* extend into Bell's Pt-zone; these are *Neuropteris tenuifolia* (in the Stubbart seam), *Sphenophyllum cuneifolium* (in the Phalen and Harbour seams), *Hymenotheca dathei* (in the Stubbart seam), and others.

Gothan (1953) asserted that *Lonchopteris eschweileriana* Andrae is a typical indicator for Westphalian A and B in Europe and thus an important fossil in the determination of the lower upper boundary of the Upper Carboniferous. Specimens of this fossil from Sydney are the wide-mesh variety and the true *Lonchopteris* is still not reported from North America. Insufficient knowledge of the distribution of this species in the Sydney coalfield opens doors for conjecture on the boundaries of the Upper Carboniferous. Thus the position of the Morien strata, as determined by fossil plants, in the Westphalian stage is by no means clear.

The concept of guide fossils is certainly useful in paleobotany, but is subject to collector's luck. To maintain viability of the concept of guide fossils seems difficult without a more rigorous definition. To inject new vigour into the concept and to avoid definitions such as "*Guide fossils remain as such as long as they have not been discovered where they are not supposed to exist*" (Arnold, 1948, p. 370), requires calculations of mathematical probability, together with a high density of sampled populations within a lithological horizon. The use of this mathematical tool ultimately leads to multivariate statistical techniques. Lyell applied a simple version of this statistical technique to the divisions of the Tertiary Basin near Paris, resulting in the well known epochs called Pliocene, Miocene, Oligocene, Eocene and Paleocene.

In most cases figured fern-like foliage in this catalogue is represented by pinnae of varying length, generally not over 28 cm (F-549), and fragments of figured fronds not exceeding, say, 20 cm. These linear measurements do not necessarily indicate the size of the structure bearing the pinnae and fronds. Fig. 1 represents a scale drawing of a petiole *in situ* in the roof of the Emery seam. By linear extrapolation, the entire structure could easily have been 160 cm (= 5 feet). This particular stalk is a specimen of *Neuropteris rarinervis*, belonging to the tree fern group (see Group 1, p. 9). A feature worth noting in Fig. 1 is the variability of the branching pattern of vascular strands in pinnules over the entire branch: younger pinnules, or terminal parts, have a simple pattern (compare to Fig. 5). In addition, terminals of vascular strands thicken. Apparently this has not previously been observed in specimens from Sydney Coalfield. The function of the thickened vascular strands is not quite clear, but it may be related to reproduction; see *Eoangiopteris andrewsii* for further discussion.

The use by Bell of Lyell's term "group" in conjunction with the terms "systems" and "stages" may be confusing (Bell, 1962a, p. viii; Bell, 1966). Group, a rock unit without any time connotation, is incorporated in his fossil listings and it appears under time-related statements (time-rock unit: relative time of the type "older than"); for additional comments see Blackadar *et. al.* (1975, p. 124-136, particularly p. 125, Article 4(d), and p. 135, Article 32(d)). Specifically, it is recorded (in Moore *et al.*, 1944, p. 679-680) that

> "Although divisions of the Pennsylvanian rocks of Eastern Canada as indicated on the chart [i.e., opp. p. 706] seem to correspond well to certain segments of geological time, the term "series", used in earlier reports, has been discarded by the GEOLOGICAL SURVEY OF CANADA [Bell, so it reads] in favour of "group", inasmuch as the use "series" is restricted to time-rock units having no geographic connotation."

"Earlier reports" refer for example to Geology Map 361A in which the concept of series is applied to the Morien and Canso type of rocks.

Identity and synonymy of species within basins and between continents

It is often observed on fronds, *N. heterophylla* for example, that differences exist between pinnules near a rachis and terminal pinnules; generally a measurable distance separates these differently situated pinnules. The gradual transition between these pinnules along a rachis is the result of ontogeny of an organ in determinate growth, and we may speak of "botanical distance" in this context. Pinnules or pinnae, therefore, that originated from different parts of one frond, now collected as fragments, may be consolidated into one species by invoking the concept of "botanical distance". However, a necessary condition for practical use of this concept is to find one complete, well-preserved frond on one whole

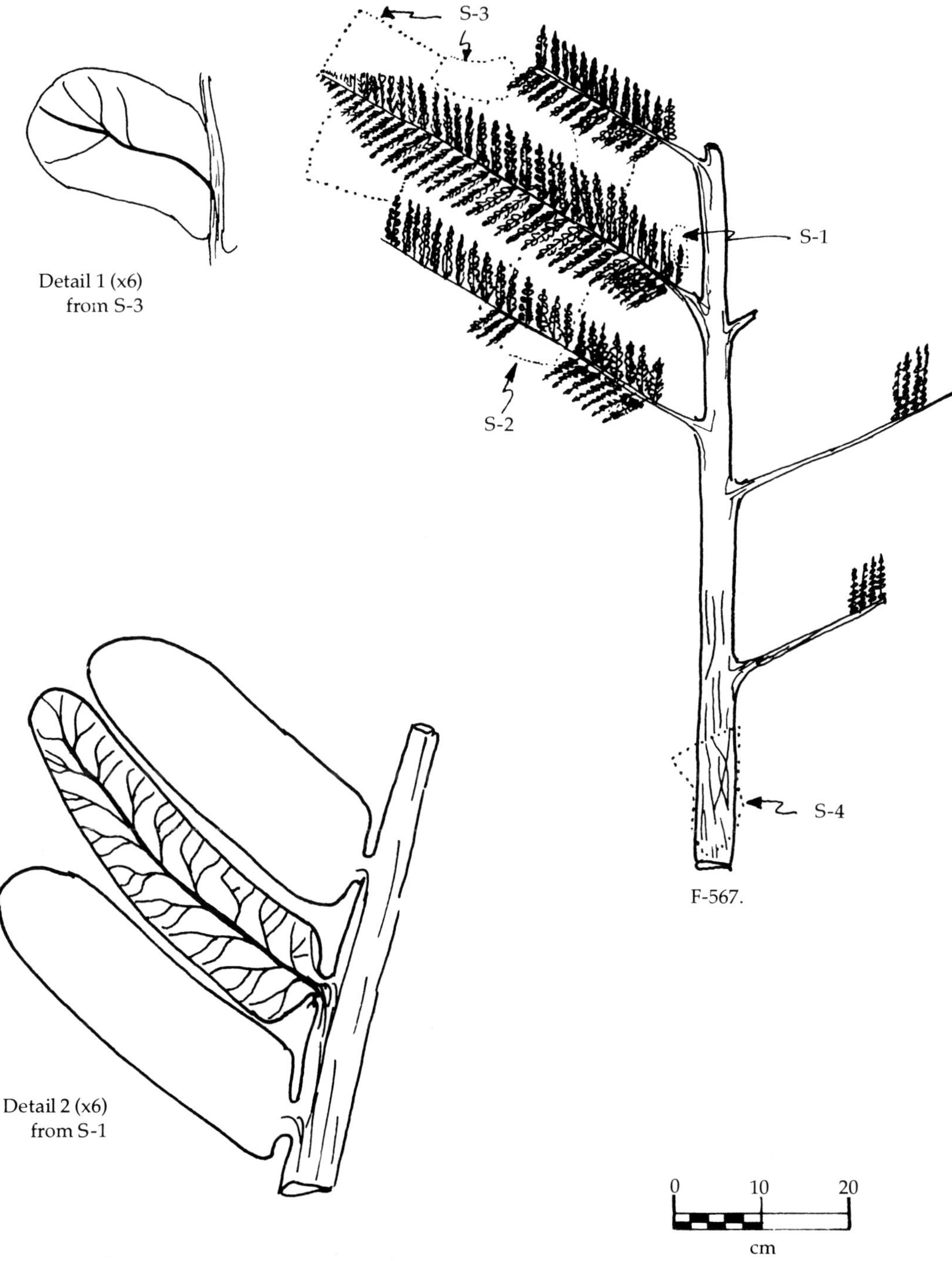

FIGURE 1
Neuropteris rarinervis: a frond of 82 cm length, F-567. This frond was located 60 cm stratigraphically above the Emery seam, 22 metres inside the bootleg* pit at Glace Bay, N.S. Areas defined by dotting in the scale drawing represent fragments of that frond in the collection: S-1 to S-4.
Detail 1 shows a pinnule from a position 5 cm below the terminal parts of the frond, S-3, and close to the terminal parts of the pinna. Note the broad attachment of this pinnule to the rachis and the simple arrangement of vascular strands.
Detail 2 shows several attached pinnules, one of which shows the characteristic venation schema of the species (from sample site S-1). The branching pattern of vascular strands in pinnules on this frond is dependent on position beginning with a simple pattern as in detail 1, dichotomizing generally in the center of the branches and then trichotomizing, as in detail 2, S-1 (compare to Fig. 4).

*An illegal mining operation.

FIGURE 2
Pinnule morphology of fern-like Paleozoic fronds. a, *Neuropteris (Mixoneura) flexuosa* (note bent veins). b, *Neuropteris*. c, well spaced vascular strands of *Neuropteris rarinervis*. Its position is two pairs of pinnules below the terminal pinnule of a pinna which is eight pinnae below the terminal parts of the frond; compare to Fig. 6 (f and g). d, the mixoneurid condition in *Neuropteris (Mixoneura)* species. e, *Allioopteris*. f, falcate pinnules having obtuse apices of *Linopteris obliqua*. g, anastomosing venation patterns of *Linopteris obliqua* (inset shows location of the enlargement). h, lanceolate pinnule showing, in part, the venation schema of *Linopteris bunburii*.

Sources: b and e, Arnold (1947, p. 161).
h is holotype 2762 (Bell, 1938, p. 65).
a, c, d, f and g, present collection.

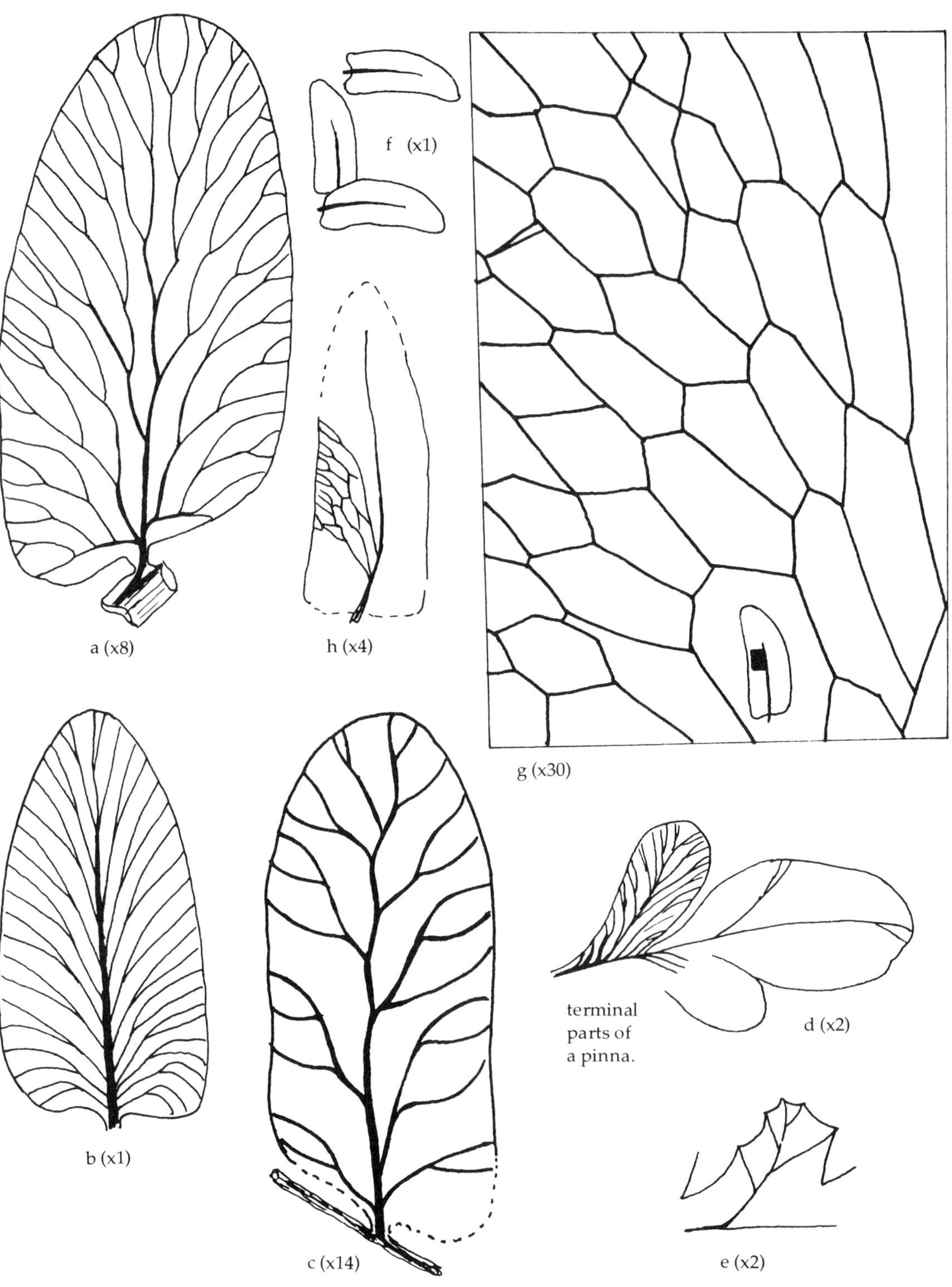

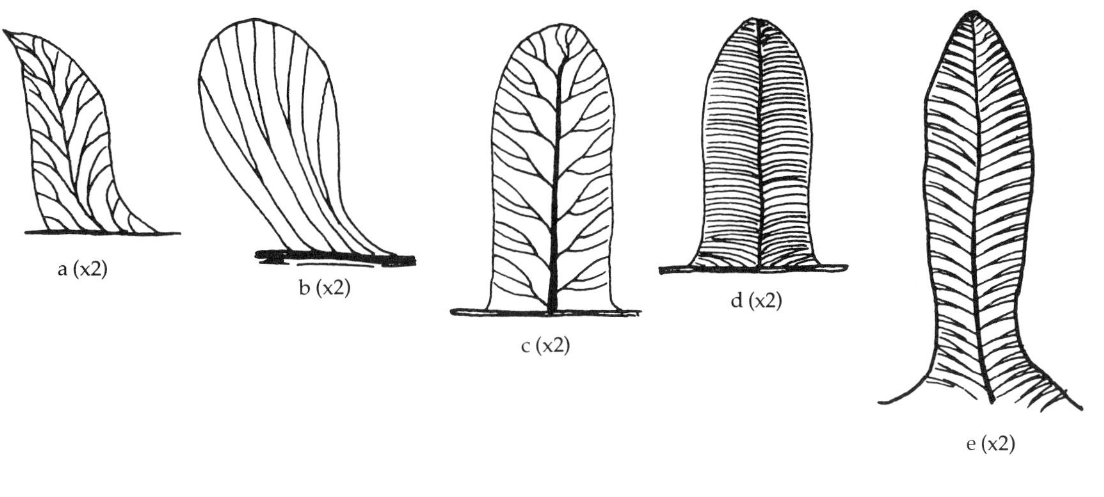

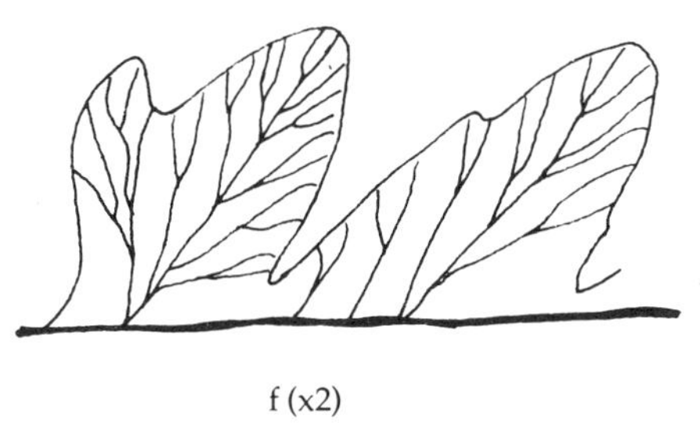

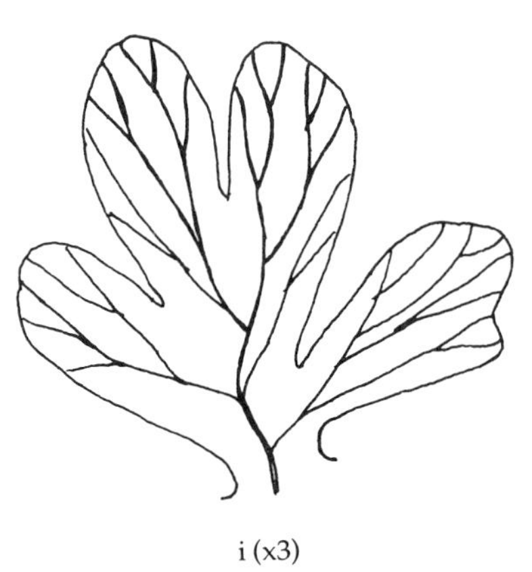

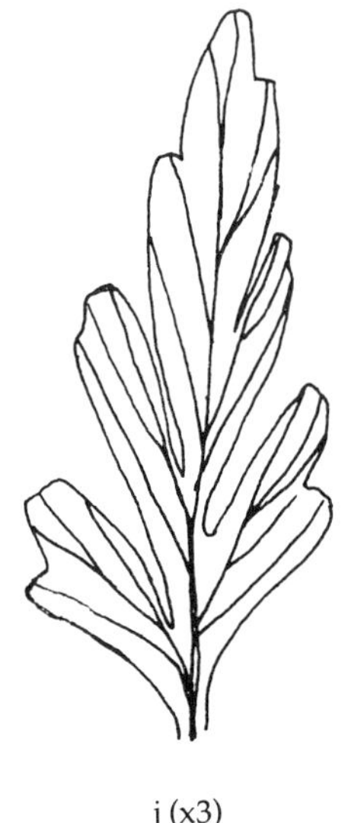

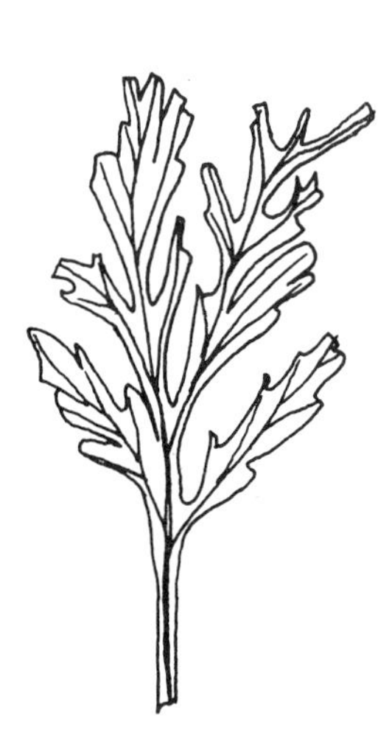

FIGURE 3
Pinnule morphology of fern-like Paleozoic fronds. a and b, *Odontopteris*. c and d, *Pecopteris*. e, *Alethopteris*. f and g, *Mariopteris*, of which g is from the Lower Pennsylvanian, Appalachian region. h and i, *Sphenopteris* (note lobate pinnule having curved vascular strands extending to margins). j and k, *Eremopteris* (note straight vascular strands and longish pinnules).

Sources: a, b, c, d, e, f and i (Arnold, 1947, p. 161).
g and k (White, 1943, Pl. 14, Fig. 7; Pl. 25, Fig. 4).
h and j from present collection.

slab of rock. This is the ideal situation in which to demonstrate identity on the species level, and is limited only by the fact that some fern fronds may be large structures (Fig. 1); refer to page 35 for a different *modus operandi.* In the case of fern fossil material from Sydney Coalfield, the large structures are fragmented by modern mining methods and sea-cliff erosion makes it virtually impossible to collect large slabs of rocks on the surface. This is a summary of one problem that faces researchers who study the synonymy of species in fossil flora within continental basins of deposition.

The next step is to study synonymy of species *between* basins of one continent in order to correlate sedimentary sequences over distances (i.e., the demonstration of time-equivalences in sequences of sedimentary deposition).

However, to study synonymy or identity of species across larger distances, or on an intercontinental scale, requires a different geologic framework and new concepts of the structural evolution of the globe. Bell (and undoubtedly his contemporary paleobotanists) used the framework of uniform organic evolution to correlate European (British and German) with Cape Breton sedimentary sequences of Upper Carboniferous age (Pennsylvanian), and Cape Breton's with New Brunswick's (Bell, 1938, p. 12-19; Bell, 1962, respectively). Brongniart's careful work and ecological reconstruction of the Carboniferous showed that the surface of the earth was hotter then, which fitted well the geophysical model of the earth at that time, and certainly influenced twentieth century paleobotanists of the pre-plate tectonic era to extend the identity of fossils on the species level from Europe to North America.

Plate tectonics, a unifying theory that brings the origins of earthquakes and geosynclines, of transgressions and regressions of oceans on continents in geologic time and other global geologic phenomena under one "umbrella", also is consistent with Brongniart's ecological suggestion for the Carboniferous period. By the theory of plate tectonics, intercontinental studies in synonymy (or identity) of fern-like, fern-type and other fossil plants of Carboniferous age may be reduced to interbasin studies on one continent. It is suggested that sedimentary basins of Carboniferous age, of which the Sydney basin with rocks of the Morien series is a part, were formed on a super-continent similar to Wegener's (1929) Pangaea and aligned with the equator. Paleobotanists now have a basis in the structure of the globe to assume even greater measures of identity and apply synonymy to species across oceans, especially the Atlantic, than ever before. The fact that the theory of plate tectonics can explain such diverse geological phenomena, with resultant ramifications in paleobotany, does not solve the problem of synonymy in a basin. Plate tectonics provides a new impetus to paleobotany and lays a firm foundation for re-interpretation, at least of Carboniferous fossils, across the present oceans.

Alphabetical Arrangement of Species

Species in this catalogue are arranged in alphabetical order with subgeneric taxa of the sort *(Asterotheca)* or *(Mixoneura)* not considered in the ordering process. In particular, the

Table 8

Synopsis: Genera in alphabetical arrangement in this catalogue, subordinated to families. (Note: Genera appearing in APPENDIX II, St. Francis Xavier University Collection, and in Addendum to INDEX OF BOTANICAL SPECIES, p. 269-273, are not listed here).

Group:	Family	Genus and its citation		
Pteridophylls	Alethopterides	*Alethopteris* Sternberg, 1825;	*Lonchopteris* Brongniart, 1836;	
	Neuropterides	*Cyclopteris* Brongniart, 1830;	*Linopteris* Presl, 1838;	*Neuropteris* (Brongniart) Sternberg, 1825;
	Palaeopterides	*Adiantites* Goeppert, 1836;	*Eremopteris* Schimper, 1869;	
	Pecopterides	*Callipteridium* C. E. Weiss, 1870;	*Mariopteris* Zeiller, 1879;	*Odontopteris* Brongniart, 1825;
		Pecopteris (Brongniart) Sternberg, 1825;		
	Sphenopterides	*Sphenopteris* (Brongniart) Sternberg, 1825;		
Pteropsids	Pecopterides	*Asterotheca* Presl, 1845;	*Eupecopteris* Kidston, 1925;	*Pecopteris* auth. cit.
		Ptychocarpus C. E. Weiss, 1869;	*Eoangiopteris* Mamay, 1950*;	*Senftenbergia* Corda, 1845;
	Sphenopterides	*Oligocarpia* Goeppert 1841, emend. M. L. Abbott, 1954;		*Sphenopteris* auth. cit.
	Marattiaceae	*Psaronius* Cotta, 1932;		
	Pecopterides***	*Alethopteris* auth. cit.	*Dicksonites* Sterzel, 1881;	*Lyginopteris* H. Potonié, 1899;
	Sphenopterides***	*Sphenopteris* auth. cit.;		
Sphenopsids	Calamariaceae	*Annularia* Sternberg, 1822;	*Asterophyllites* Brongniart, 1828;	*Calamites* Brongniart, 1828, ICBN, p. 376**
		Macrostachya Schimper, 1869;	*Palaeostachya* C. E. Weiss, 1876;	
	Sphenophyllaceae	*Sphenophyllum* Brongniart, 1828: ICBN, p. 376.**		
Lycopsids	Lepidodendraceae	*Lepidodendron* Sternberg, 1820;	*Lepidodendropsis* Lutz, 1933;	
		Lepidophloios Sternberg, 1825;	*Lepidophyllum* Brongniart, 1828;	
		Lepidostrobophyllum Hirmer, 1927;	*Lepidostrobus* Brongniart, 1828;	
	Sigillariaceae	*Sigillaria* Brongniart, 1822;	*Sigillariophyllum* Grand'Eury, 1877;	
		Sigillariostrobus (Schimper) Eugen Geinitz, 1873;	*Stigmaria* Brongniart, 1822;	
Pteropsids	Cordaitaceae	*Artisia* Sternberg, 1838;	*Cordaianthus* Grand'Eury, 1877, ICBN, p.377;**	
		Cordaites Unger, 1850;		
	Unknown affinity	*Triletes* Reinsch, 1881.		

*Indexed as part of *Pecopteris-Ptychocarpus.* **Stafleu *et al.*, 1972. ***Class Gymnospermae.

Source: Andrews (1955).

species are arranged according to Tables 1, 2, 3, or 5 with genera as the primary alphabetical ordering under the appropriate families. The sequence of families in the catalogue is arbitrary. The schema of ordering is summarized in Table 8.

In this catalogue species are subordinated to the Linnaean hierarchy. The wisdom of this may be questioned in the light of the many uncertainties surrounding fossil plants, their interrelation in Carboniferous time, and their relation to present flora. However, justification or defense is simply based on the desire for order, no matter how artificial it appears at this time, or how controversial. The alternative is "nomenclatural anarchy".

Note on Infrared Photography

Some use is made of infrared reflection photography in this catalogue (high speed infrared film on dimensionally stable Estar base, Kodak Publication, 1974). The authors believe that this technique has potential in photographing fossils.

First, this type of photography under certain conditions allows an "inside" look into coalified layers of fossil material. For example, in *Eupecopteris (Asterotheca) cyathea,* Pl. 73, Fig. 1, black and white photography indicates by shadows the structures on the surface of pinnules; it is infrared reflection which in this case shows the inner structures resembling spore cases. Also of interest are the infrared renditions of *Asterotheca daubreei* and *Asterotheca herdi.* In the first case infrared images show structures resembling, in a superficial way, a restored *Crossotheca* sp. (Andrews, 1961, p. 136, Fig. 5-4), while in the second case infrared images of the *A. herdi* specimens did not improve photographic contrast. Both fossils are embedded in dark, micaceous, fine-grained sandstone (Plates 65 and 68).

Infrared reflection photography is also useful in situations where carbon compressions show fine venation detail and the host rock is not coated with ferrous oxide. A good example is the gray shale and its fossils from the Stubbart seam at Prince Mine, Point Aconi. *Neuropteris flexuosa* specimens, preserved as carbon compressions, render black and white images useless because of reflection; polarizing filters on the camera obscure effectively the venation pattern but reduce glare. Infrared reflection images, however, show excellent contrast and enhance visualization of venation, as shown on Plate 24, Fig. 2.

However, low resolution was obtained from infrared images of carbon compressions of a specimen of *Linopteris obliqua* embedded in shale that has extensive surfacial ferrous oxidation, although the specimen possesses good venation detail. Bell is the determiner of the species, Appendix II; see Plate 20.

All cases discussed are illustrated in appropriate plates of the catalogue and all additional infrared photography is explicitly identified.

Explanations for the Enumeration of Species in this Catalogue

Criteria for fossil classification and nomenclature

PTERIDOPHYLLEAE: (class according to Table 1).

Neuropterides: (family according to Table 1). It should be remembered that Table 1 includes both pteridosperms and true ferns; it is our ignorance that prevents resolution.

Filicites tenuifolius Schlotheim: a genus and species under the class and family, with Schlotheim the author of both the trivial name *tenuifolius* and the name of the genus (von Schlotheim, 1820, p. 405, Tafel XXII, Fig. 1).[5] Brongniart, after having recognized that this

5 Schlotheim's use of *Filicites* rendered a type species meaningless because of the diversity of fossils assigned to the genus (Andrews, 1955, p. 158).

species is encompassed by the form-genus *Neuropteris,* transferred Schlotheim's species to:

(1) *Neuropteris tenuifolia.*

In effect, Brongniart placed Schlotheim's species in another genus and for the first time gave it a meaningful status (Brongniart, 1829, p. 241-242, Pl. 72, Fig. 3, 3A). Paleobotanists now cite Schlotheim's species as:

(2) *Neuropteris*	*tenuifolia*	(Schlotheim)	Brongniart,	1829[6]
genus	species	author of species *F. tenuifolius*	changer of genus from *F.* to *Neuropteris*	year generic status changed

6 Recommended by the ICBN, i.e., the year of change of the genus should be included.

thus crediting Brongniart, in the literature, with that change of genus. The parentheses (the "bracket convention") enclosing "Schlotheim", followed by the second name, "Brongniart" in this case, mean as a general rule that someone a) transferred a species to another genus, or b) placed the species under another generic name for the same genus without a change in rank. The former case applies to (2) above. For detailed information and additional examples, the reader is referred to "International Code of Botanical Nomenclature" (ICBN), 1972, Article 55 (Stafleu *et al.*, 1972, p. 53).

Author/changer relationships also exist in generic names; the form-genus *Neuropteris* as used in (1), first employed by Brongniart as a family name, should be cited as:

(3) *Neuropteris*	(Brongniart)	Sternberg,	1825
genus	author of basionym	author of new combination	year of combination

7 However, see REMARKS under this fossil listing in the catalogue section.

(with associated type species *Neuropteris heterophylla* (Brongniart) Sternberg, 1825. See also Table 8). Brongniart used "Nevropteris" as a subgenus of *Filicites:*

(4) *Filicites (Nevropteris) heterophyllus* Brongniart, 1822

Sternberg in 1825 gave it generic rank for the first time and changed the "v" to a "u" *(Neuropteris),* (Note: the dates of changes in the third and fourth examples are from Andrews (1955, p. 195). For details of nomenclatural rules see ICBN (Stafleu *et al.*, 1972, p. 48, Article 49). For an example of a homonym refer to ICBN (Stafleu *et al.*, 1972, p. 376). *Calamites* Brongniart, 1828 with type species *C. radiatus* Brongniart is a *nominum conservandum,* while *Calamitis* Sternberg, 1820 with type species *Calamitis pseudobambusia* Sternberg is a *nominum rejiciendum.*).

It can be appreciated that the bracket convention may become a problem in paleobotanical nomenclature, confusing students and experts alike.

Bell (1938), in his important contribution to knowledge of fossil flora, studied and figured Upper Carboniferous fossil plants from Sydney Coalfield. These fossils, among others, are listed in his catalogue (1962a). The species in (2), p. 21, is listed as:

(5) *Neuropteris tenuifolia* (Schlotheim).

The round parentheses not being followed by the changer's name is an application of Article 51(d) of the International Code of Zoological Nomenclature (Stoll *et al.*, 1961, p. 51):

"Example . . .

(i) The use of parentheses here applies only to transfers from one nominal genus to another, and is not affected by the presence of a subgeneric name, or by any shifts of rank or position within the same genus."

Use of "cf.", "sp. indet.", "sp.", "gen. indet." and references with botanical names

If specimens are imperfectly preserved or lack sufficient diagnostic features for specific identification then a new specific name may not be warranted; this is indicated by "cf." *(conformis)* before the trivial name, or before a botanical name in a complex situation such as this:

Sphenopteris sp. Bell cf. *Sphenopteris dufayi* Danzé.

Here, the *Sphenopteris* species described by Bell (in a scientific journal) is similar to Danzé's species *Sphenopteris dufayi* (described before Bell's in a scientific journal), but Bell in this case had doubts about the specific identity of his specimen because of lack of diagnostic features. The "cf." designation implies that the given genus is not disputable, but placement into species is not feasible with any accuracy (Orlov, 1962, p. 103-4). Thus "cf." must be followed by a trivial name. The undeterminable species is indicated by "sp. indet." *(indeterminatum);* at times we use "sp." (instead of "sp. indet.") to indicate this condition. However, as in the "cf." case, the genus associated with "sp. indet." and "sp." is not disputed. The reasons for indeterminacy of species are varied; in many cases it is because of poor fossil preservation which may mean lack of sufficient diagnostic features. The line of demarcation between "cf." and "sp. indet.", with all its implications, depends naturally on the individual researcher, and varies among paleobotanists (Dr. Arnold was considered to be conservative in his determinations).

Asterotheca cf. *abbreviata* Brongniart: [Crookall, 1929, Pl. XXVI], means that Crookall is a listed reference for this species.

Sample Data and Ranges of Fossils (see Table 7)

F-288, F-105, 967G10.1 means that these specimens are in the collection; F-215-217 means the figured specimens are located in the same bedding plane of one block.

FOSSIL ASSOCIATIONS:
A general observation indicating that listed specimens are in the same bedding plane, or nearly so, of one block. "Nearly" means within one to two centimetres. "Fossil association" is used in preference to "sympatry".

SAMPLE LOCATIONS:
For exact geographic and stratigraphic location of specimens in the collection consult GEOGRAPHIC AND STRATIGRAPHIC LOCATIONS OF SPECIMENS F-1 to F-682.

However, "Tracy seam", "Emery seam", and so on imply without dispute that the sample originated above those seams in the stratigraphic sense. Refer to Table 6.

BELL'S RANGE:
Refers to species collected in Sydney Coalfield only and indicates the ranges of species according to Bell (1938, Fig. 1). See Table 6. The citation of Bell's range in the Catalogue of Species is maintained.

Nomenclature and Synonymy in this Catalogue

In this catalogue the practice of (5) on p. 22, if it applies, is followed consistently not only to save space but also to reduce the possibility of typographical errors. The application of this rule does not affect the possibility that author and changer are one and the same person, as in:

(6) *Sphenophyllum emarginatum* (A. T. Brongniart) A. T. Brongniart.

This is an infrequent case (Stafleu *et al.*, 1972, p. 376). In the INDEX TO BOTANICAL SPECIES an attempt is made at complete citation.[8] Identity of species is referenced in this INDEX thus:

Alethopteris friedeli = *Alethopteris davreuxi*,

where *A. friedeli* is the invalid name (to be replaced by *A. davreuxi*). The primary entry in the catalogue, however, of this collected species is under *A. davreuxi* with a reference to the invalid name under REMARKS; both names are cross-referenced in the INDEX. In the INDEX the invalid name of a species, "*(A. friedeli)*" for instance, is indented directly beneath the valid name (note the arrangement of parentheses). In addition, indented

8 This may not apply to species in Appendix II, which are specimens from the collection in the Dept. of Geology, St. Francis Xavier University.

names of species *without* parentheses appear in the INDEX directly below a name; this arrangement signifies similarity to, or variety or forms of, or is a reference to the species directly above:

	page number
Alethopteris valida Boulay	**30, 107**
Validopteris integra Gothan	30

means that a specimen indexed under *A. valida* bears resemblance to *V. integra;* hence *V. integra* appears under *A. valida,* the primary indexed species.

Readers are referred to Dijkstra and to Jongmans for general citation of species (genus + species + author, see (2) on p. 21) and to Bell (1966, 1962a, 1962, and 1938) for references on conspecifics, synonymy and homonymy of species from Sydney Coalfield.

Acknowledgements

The present paleobotanical investigation was made possible by grants from the following sources:

The Council for Evaluation of Research Proposals, College of Cape Breton, Sydney, N.S.;

Nova Scotia Museum, Halifax, N.S., Mr. J. L. Martin, Director;

Bras d'Or Institute of the College of Cape Breton, Sydney, N.S., Professor D. Arseneau, Director.

The authors are pleased and very grateful to Mrs. D. M. Sutherland, Head Librarian, Geological Survey of Canada Information Processing Division, for having made available to this investigation the monumental works by R. Kidston and R. Crookall, and other material. Officials of the Coal Division of the Cape Breton Development Corporation in Sydney deserve thanks for their enthusiastic response to requests for underground and mine dump visitations. They are:

Mr. J. Bardswich, Vice-President,
Mr. G. MacLean, Manager, Administration,
Mr. A. R. MacLean, Director of Mineral Resource Planning,
Mr. B. Phillips, former Manager, Lingan Mine; S. Farrel, now Superintendent of Mines,
Mr. D. Merner, Manager, Prince Mine at Point Aconi,
Mr. John T. LeBlanc, former Manager, Princess Mine, and
Mr. J. MacLellan, Manager of #26 Mine.

We also thank the coal miners at Lingan who made valuable comments on fossil locations in that mine. We acknowledge permission from Dr. W. S. Shaw, Geology Department, St. Francis Xavier University, Antigonish, Nova Scotia, to include its collection in this catalogue.

The College of Cape Breton and the Nova Scotia Museum are proud of their association with Dr. Chester A. Arnold,* Emeritus Professor of Geology and Botany and Emeritus

*Deceased Nov. 19, 1977.

Curator of Paleobotany, the University of Michigan Paleontological Museum, Ann Arbor, Michigan. As visiting scientist at the College and the Nova Scotia Museum, Dr. Arnold performed taxonomic work to the species level on fern-like foliage and other fossil plants of Carboniferous age from Nova Scotia. This is by no means an easy undertaking if remarks by H. P. Banks, Cornell University, are considered (*in litt.*, July 21, 1975):

> "You [Prof. Zodrow] can be sure that few individuals in North America can or will identify Carboniferous plants at the species level."

We are indebted to Dr. Arnold for valuable discussions and liberal use of one chapter (Pteridophyllophyta) of his MS textbook on fossil ferns, and to Dr. J. Reynolds, University of New Brunswick, for valuable nomenclatural suggestions and criticisms.

In addition, D. A. Reeve, Assistant Head of the Canadian Metallurgical Fuel Research Laboratory at Ottawa, deserves thanks for X-ray powder diffraction and optical analyses of coal-associated minerals, and Dr. L. J. Cabri, Head, Mineralogy Section, Canadian Centre for Mineral and Energy Technology, Department of Mines and Resources Canada, for the identification of the minerals melanterite and rozenite from the Phalen seam at #26 Colliery.

The authors are indebted to Dr. T. E. Bolton, Curator of Type Collections of the Geological Survey of Canada, for receipt of original photographs by Bell of his *Linopteris obliqua* var. *bunburii* Bell (Plate 21, Figs. 2 and 4; Plate 22, Fig. 1) and nomenclatural suggestions; to Dr. W. N. Stewart and J. Doran, Department of Botany of the University of Alberta, and Dr. D. C. McGregor, Geological Survey of Canada, for nomenclatural suggestions. Dr. Stewart, in addition, made the following specific determinations:

F-190, *Sphenopteris* cf. *hoeninghausi;* F-174, *Oligocarpia* cf. *missouriensis;* F-412, *Sphenopteris* cf. *suspecta;* F-68, F-71, F-78, *Neuropteris heterophylla;* F-101, *Sphenopteris* cf. *striata;* F-237, *Ptychocarpus unitus;* F-97, *Lycopodites meekii* (refer to *Lepidodendron dawsoni);* F-86, *Neuropteris rarinervis;* F-173, *Senftenbergia* sp.; and Block F-248, *Pecopteris (Asterotheca) acadica, Asterophyllites equisetiformis* and F-737, *Trigonocarpus* species.

Ausserdem sind wir Herrn Zeschke, Institut für Geologie und Paläontologie, Technische Universität in Berlin, Dank schuldig für geliehende Beihefte Gothans (1953) und Dabers (1955), sowie für den Hinweis auf Weylands neuste bearbeitete Ausgabe des Lehrbuch Gothans (1973).

Triletes auritus var. *grandis,* a tentative identification, was made by Mr. M. Sedley Barss, Atlantic Geoscience Center, Geological Survey of Canada, Dartmouth, N.S. We appreciate his expertise.

We owe special thanks to Mrs. Hilda Day of the Beaton Institute of the College of Cape Breton for having made available to us (on a permanent loan basis) the following fossils:

F-497, *Calamites suckowi;* F-501, *Neuropteris scheuchzeri* (an exceptional specimen); and F-499, F-502, *Pecopteris obtusa.* The last named species is listed under Addendum to INDEX OF BOTANICAL SPECIES.

Mrs. Day was given these specimens by Miss Ethel Johnstone (1892-1974) who was a great-granddaughter of R. Brown; see entry under *Linopteris obliqua* for a short note on the late R. Brown of Sydney, N.S.

In addition we thank Dr. D. S. Davis, Chief Curator of Science and Mr. R. Grantham, Curator of Geology, Science Section of the Nova Scotia Museum in Halifax, for use of equipment, their hospitality, but most of all for their enthusiasm for this project. We also thank Mr. Niels Jannasch for his careful translation from E. F. von Schlotheim used in the Foreword.

The authors are thankful to Dr. D. F. Campbell, President of the College of Cape Breton, for his continuing support of research efforts at this College through the establishment of The Committee for Evaluation of Research Proposals, chaired by the Academic Vice-President Dr. W. Reid. It is CERP that unflinchingly provided continuous funds for this project which we interpreted as encouragement of our activity.

The authors owe a special debt to Mrs. Barbara Shaw and the editorial board of the Nova Scotia Museum; specifically we mention Mr. F. Scott, Curatorial Assistant (Science Section), for his invaluable editorial efforts, and time spent on the manuscript to improve style, point out conflicts and the like. We thank him. To my wife, Kris Zodrow, I owe thanks for special efforts in proof-reading this manuscript.

Catalogue of Species

PTERIDOPHYLLEAE[1]

ALETHOPTERIDES[2]

Alethopteris davreuxi (Brongniart)[3] (F-4, F-41, F-57B, c, F-92b, F-210-14, F-272, F-329a, F-335, F-337, F-361-64, F-360, F-522, 966G57.7, 967G16.13, 967G16.15, 967G36.4).

Plates 1, 2; Plate 3, Fig. 3

FOSSIL ASSOCIATIONS:
F-210-11: with *L. obliqua* (Bunbury).
F-360: with *A. valida* Boulay, F-359.
F-272: with *N. (Mixoneura) flexuosa* Sternberg, F-266-68; *N. flexuosa* forma *magna*, F-270; *Palaeostachya* sp. indet., F-269, F-271; and *L. obliqua* (Bunbury) and *Sphenophyllum* sp. indet.

SAMPLE LOCATIONS:
F-57b, c: Mc Aulay seam.
F-92b: Shoemaker seam.
F-522: Emery seam.
F-41: Harbour seam, #12 Mine dump.
F-329a, F-335, F-337, F-361-64, F-360: Harbour seam Lingan Mine dump.
16.13, 16.15[4]: Harbour seam, Florence Mine.
F-210-214, F-272: Stubbart seam, Prince Mine.
F-4: Unknown location in the Morien series, Cape Breton Island.
57.7: Morien series?
36.4: Unknown, assumed to be from Nova Scotia.

BELL'S RANGE: L-Pt-zones.[5]

REMARKS:
In addition, *L. obliqua* occurs with *A. davreuxi* and *N. (Mixoneura) flexuosa* on block F-255. With the exception of F-57b, c, which are detached pinnules, the remaining specimens are detached pinnae, some with terminal pinnules (F-4, F-92b, F-362-64). *Alethopteris friedeli* P.

1 Refers to class, as in Table 1.

2 Refers to family, as in Table 1.

3 For the significance of round parentheses in "(Brongniart)" refer to p. 22.

4 Accession number is shortened, at times also in REMARKS.

5 Maintained for historical reference, see SAMPLE DATA AND RANGES OF FOSSILS.

Bertrand is a synonym of *A. davreuxi;* see Bell (1962, p. 37). A general characteristic of this family is possession of pinnules which are broadly attached to the rachis. The pinnules may be parallel-sided or triangular with scalloped (rarely serrated) margins and distinct midribs. All of the lateral veins originate from the midrib (which extends nearly to the apex). The laterals are simple or forked.

Many pecopterid fronds bear fern-type fructifications, for example *Asterotheca, Cyathotrachus, Ptychocarpus* and *Renaultia,* among others (see Table 2). It is of interest to note that some pecopterids, of which *Pecopteris pluckeneti* is a good example, were found to be seed-bearing.

Note on Taxonomy: We call attention to the possibility that Goeppert may not have emended the genus.

Alethopteris decurrens (Artis) (976GF22.1).

Plate 3, Figs. 1, 2

SAMPLE LOCATION:
Unknown, assumed to be in Nova Scotia.

REMARKS:
A well-preserved 6 cm pinna.

▶ *Alethopteris grandini* Brongniart (967G123.18).

Plate 4, Fig. 1

SAMPLE LOCATION:
Morien series, Cape Breton Island.

REMARKS:
A 10 cm partial pinna with terminal parts missing.
Not previously reported from Sydney Coalfield.

Alethopteris lonchitica (Schlotheim) (967G10.72).

Plate 4, Fig. 3

SAMPLE LOCATION:
Morien series, Cape Breton Island.

BELL'S RANGE:
Lch-zone (Mc Aulay seam).

REMARKS:
The specimen is a pinna of 5 to 8 cm which shows terminal parts. There is a tendency in European literature to lump this species and *A. serli* together as *A. lonchitica* Schlotheim, including forma *serli* Brongniart (Gothan, 1953, p. 16; Gothan and Weyland, 1973, p. 299).

We have evidence (F-595(6) verso) to suggest that this species was present in the roof of the Emery seam, Glace Bay.

Alethopteris scalariformis Bell (967G10.73, 967G123.22).

Plate 4, Fig. 2; Plate 5

SAMPLE LOCATION:
Morien series? Cape Breton Island.

BELL'S RANGE: L-Pt-zones.

REMARKS:
Specimens consist of terminal parts of pinnae, the largest of which is 11 cm in specimen 10.73.

Alethopteris serli (Brongniart) (F-140, F-143, F-144-145, F-149, F-150, F-154, F-388, F-389, F-568, 967G16.14, F-574, F-594, F-618).

Plate 6; Plate 7, Figs. 1,2; Plate 8, Figs. 2-4; Plate 9, Figs. 1, 3

FOSSIL ASSOCIATIONS:
F-143: with *Neuropteris (Mixoneura) ovata* Hoffmann.

F-150: with *N. scheuchzeri* Hoffmann; see also F-594 for similar association.
F-618: with *Pecopteris (Senftenbergia)* cf. *pennaeformis* Brongniart.

SAMPLE LOCATIONS:
All specimens are from the roof of the Emery seam, Glace Bay, except for 16.14 (from the Florence Mine, Harbour seam) and F-618 (from Stubbart seam, Prince Mine).

BELL'S RANGE: Morien series.

REMARKS:
F-140, F-149 and F-594 are good examples of terminal parts of pinnae, while F-568 represents the tip of a frond. F-154 bears a superficial resemblance to *Alethopteris valida* but it is believed to be the terminal part of a frond of *A. serli*. F-574 has attached pinnae. The Emery seam is a good location for collecting this species.

Alethopteris valida Boulay (F-93, F-359, F-452 verso, F-455, F-456, F-570, F-586 verso, 867G1.1, 912G7.1).

Plate 7, Fig. 3; Plate 8, Fig. 1; Plate 9, Fig. 2;
Plate 10

FOSSIL ASSOCIATIONS:
F-359: with *A. davreuxi* (Brongniart).
F-456: with *N. scheuchzeri* Hoffmann, similarly for F-452 verso.

SAMPLE LOCATIONS:
F-93: Shoemaker seam.
1.1: Gardiner (?) seam, Schooner Pond, Little Glace Bay.
F-586 verso: Emery seam.
F-452 verso, F-455, F-456, F-570: Phalen seam.
F-359: Harbour seam, Lingan Mine dump.
7.1: Cowbay Bar, Morien series, Cape Breton Island.

BELL'S RANGE: L-Pt-zones.

REMARKS:
F-455 seems to be an ultimate pinna of the *Validopteris integra* type (Bell, 1938, p. 68-69). Both F-570 and F-586 specimens have rather large pinnules.

Lonchopteris eschweileriana Andrae (F-94, F-390, F-460, F-461, F-462, F-463).

Plates 11, 12

FOSSIL ASSOCIATION:
F-390: with *Neuropteris scheuchzeri* Hoffmann.

SAMPLE LOCATIONS:
F-94, F-460, F-461, F-462, F-463: Shoemaker seam.
F-390: Emery seam.

BELL'S RANGE: Lch-zone.

REMARKS:
The only known locations of this species in North America are the coal seams noted above. At these localities this species is rare. F-94 is a fragment of a frond, while F-390 is a pinnule of 6 cm, probably the terminal part of a pinna(?).

F-94, F-460 and F-463 are from the upper position in a frond as indicated by the confluence of pinnules (forming secondary pinnae) which are expressed as marginal lobes. F-462 seems to have originated from the lowest position of frond F-460.

Gothan (1953, p. 28-30, 32-33) in discussing the geographic distribution of this genus noted that it is a typically Middle European form occurring in the continental basin in Northern France — Belgium —Aix-la-Chapelle (Germany) — Ruhr — Upper Silesia; in the East the genus is not recorded from the Donets Basin, and in the West, viz. England, the genus is practically absent except for Southern Wales where a few specimens are found. Lately, Bell (1938, p. 70) reported the species from Canada, but his finds represent the wide-meshed variety of the genus and the true *Lonchopteris* is still not reported from North America. Gothan asserted that the American Upper Carboniferous seems, therefore, to connect with the West European Upper Carboniferous. The true *Lonchopteris* is typical of the Upper Middle Pennsylvanian (Westfalian A and B, see Table 7) and seems an important fossil in the determination of the lower upper boundary of the Upper Carboniferous.

The uncertainty of the distribution of this species in the Sydney Coalfield opens doors for conjecture on the boundaries of the Upper Carboniferous.

Boersma (1978) in studies of the Taphoflora of the Illinger Floezzone, Germany, placed a floral assemblage homotaxial with that of the upper *Linopteris obliqua* zone (see Tables 7 and 10) in the lowermost Stephanian A.

NEUROPTERIDES

Cyclopteris cf. *fimbriata* Lesquereux (F-378, F-443).

Plate 13, Figs. 1, 2

FOSSIL ASSOCIATIONS:
F-378: with *Linopteris obliqua* (Bunbury).
F-443: with *N. scheuchzeri* Hoffmann.
Block F-513: with *Annularia stellata* Schenk.

SAMPLE LOCATIONS:
F-378: Harbour seam at Lingan Mine.
F-443: Emery seam.

BELL'S RANGE: L-Pt-zones.

REMARKS:
Attributed by Bell to *Neuropteris (Mixoneura) ovata*.
Resembles a squashed "thistle", with fimbriated borders (Bell, 1938, p. 61).

General characteristics of the family are: pinnules, with entire margin, may be rounded, oval or tapered or linguoid; attachment, although exceptions are noted, of pinnule with rachis is by downward extension of the midrib(s) of the pinnule. Of conspicuous nature is *Linopteris,* which displays anastomosis of veins.

Cyclopteris sp. indet. (F-430, 967G20.2).

Plate 14, Fig. 1

FOSSIL ASSOCIATION:
F-430: with *Neuropteris scheuchzeri* Hoffmann.

SAMPLE LOCATIONS:
F-430: Harbour seam at Lingan Mine.
20.2: Two miles east of Stellarton, N.S., at McLellan Brook, Pictou Co., Pictou series.

REMARKS:
The associated cyclopteroid pinnule of *Neuropteris scheuchzeri* measures 5 cm.

These pinnules are generally situated on the main rachis below the lowermost primary pinnae and belong to the same group as *Cyclopteris orbicularis* Brongniart.

Cyclopteris pinnules, at times with fan-shaped venation and without midribs, are found mostly detached and belong to several different pteridophyll fronds.

Linopteris muensteri (Eichwald) (F-334).

Plate 13, Fig. 3; Plate 14, Fig. 2

FOSSIL ASSOCIATIONS:
With *Alethopteris davreuxi* (Brongniart) and *Neuropteris scheuchzeri* var. *Dawsoni* (?), F-336.

SAMPLE LOCATION:
Harbour seam at Lingan Mine.

BELL'S RANGE: Lch-L-zones.

REMARKS:
The specimen is a single, detached pinnule, the only one found.

Linopteris neuropteroides (Gutbier) (F-110, F-111).

Plate 15, Fig. 1

SAMPLE LOCATION:
Tracy seam.

BELL'S RANGE: L-zone?

REMARKS:
It is not clear whether these specimens conform to var. *major* H. Potonié. (Bell, 1938, p. 66). They were collected as single detached pinnules, which makes specific identification difficult.

Linopteris obliqua (Bunbury) (L.0.-1 to 30, F-229, F-232, F-248, F-263, F-264, F-265, F-272, F-285-3,-4, F-288 verso, F-427, F-428-1, F-616, F-617, F-644, F-657).

Plate 15, Fig. 2; Plates 16-21; Plate 22, Fig. 1

FOSSIL ASSOCIATIONS:
F-263: with *Macrostachya infundibuliformis* (Brongniart), F-259.
F-265: with *Sphenophyllum emarginatum* (Brongniart).
F-272: with *Alethopteris davreuxi* (Brongniart),
Neuropteris (Mixoneura) flexuosa Sternberg, F-266,-68,
Palaeostachya sp. indet., F-269, F-271.
F-288 verso: with *Eremopteris artemisiaefolia* (Sternberg).
F-427: with *Asterotheca daubreei* Zeiller, and *N. scheuchzeri* Hoffmann.
F-616, F-617: with *Ptychocarpus unitus* (Brongniart).

SAMPLE LOCATIONS:
L.O.-1 to 30, F-288 verso, F-427, F-428-1: Harbour seam, Lingan Mine dump.
F-299, F-232, F-248, F-263, F-264, F-265, F-272, F-285-3,-4, F-616, F-617, F-644, F-657: Stubbart seam, Prince Mine, Point Aconi.

BELL'S RANGE: L-zone. Note: L-Pt-zones for *L. obliqua* var. *bunburii* Bell (= *Linopteris bunburii* Bell, 1962, p.5).

REMARKS:
We record for the first time direct associations of *L. obliqua* and *Ptychocarpus unitus* from the Sydney Coalfield (F-616, F-617). L.O.-27 is a longitudinal fragment of a pinna on which the falcate shape of the pinnules leans towards the terminal part of the pinna; F-657, a second specimen, is a fragment of a pinna; all other pinnules are detached.

L.O.-1 to 30:
All of these specimens may be regarded as *L. obliqua* (Bunbury). Bunbury's original figured specimens were from Cape Breton Island, although there is considerable resemblance to *Linopteris neuropteroides*. The latter, however, is a very rare species and one cannot be sure how readily it may be distinguished from *L. obliqua* when variations in the latter are considered. As for Bell's species, it would require Bell's judgment for their separation of the two species. Both species, *L. obliqua* and *L. bunburii* may occupy different positions on one frond (in that case they are conspecific). See Fig. 4 for variations of pecopterid venation over one frond. This species is of the par form.

Through the courtesy of T. E. Bolton, Curator of Type Collections, Geological Survey of Canada, the authors received prints of Bell's original negatives of *L. obliqua* var. *bunburii* holotype GSC 2762 and paratype GSC 7052B. Through a comparison between F-285-3 and Bell's photographs we conclude tentatively that our specimens fit *L. bunburii*.
For historical accuracy and interest, Bunbury's Figures (1847, p. 427) are reproduced in Fig. 6. He originally described it:

Spec. Char. *Dictyopteris obliqua* (n. sp.)
Pinnulis oblongis obtusissimis subfalcatis, basi oblique subcordatis; costa tenuissima; venis prominulis.

Unfortunately, Bunbury did not list sample locations for this species, or for *Neuropteris rarinervis* and *Odontopteris subcuneata,* in the Sydney Coalfield. These specimens, among others, were sent to Bunbury by the colorful and illustrious R. Brown, Esq., General Manager of the General Mining Association (of London) in Sydney, N.S. The interested reader is referred to the Beaton Institute, College of Cape Breton, Sydney, N.S., for a personal collection on R. Brown. Dawson in his *Acadian Geology*, 1878, acknowledged that his knowledge of the geology of coal and fossil plants from Cape Breton Island was based on Brown's studies. Consult Gregory (1975, p. 79-80) for Brown's publications. More recently, Milligan (1977, p. 589) emphasized the role played by R. Brown in the development of thought about the nature of coal deposits.

It appears that *Linopteris neuropteroides* Geinitz (Crookall, 1959, v. 4, Pt. 2, p. 209) and *L. neuropteroides* (Gutbier) (Gothan, 1953, p. 66) are conspecific with *L. obliqua* (Bunbury) and *L. bunburii* Bell since there are no significantly different taxonomic characteristics; this is supported by Table 9.

See Tables 9 and 10 for summaries of pinnule characteristics of some *Linopteris* species, and fossil plants that are associated with *L. obliqua*.

We have not used the name *Linopteris neuropteroides* Geinitz for three reasons (personal communication with Dr. Arnold, 1977):

1) nobody has pointed out criteria for the separation of *L. obliqua* and *L. neuropteroides* that are consistent and efficient. See Crookall (1959, v. 4, Pt. 2, p. 210-212) for a summary, and Zodrow and M^c^Candlish (1978).
2) Bunbury's species came from Cape Breton Island and descriptions and other material were published in 1847.
3) Geinitz published his material and description of his species in Germany in 1855.

Table 9

Summary of comparison of pinnule morphology of some *Linopteris* species from Sydney Coalfield and elsewhere.

Pinnule Morphology	*Linopteris obliqua*: Cape Breton specimens Bunbury (1)	Bell (2)	Arnold (3)	*L. bunburii* Bell (2)	*Linopteris neuropteroides* Gutbier (4)
Attachment	all species are point-attached				
Length (5)	≃ 11 to 23 mm	≃ 14 to 25 mm	≃ 10 to 30 mm	≃ 10 to 17 mm	≃ 33 mm
Width (5)	≃ 5 to 7 mm	≃ 5.5 to 8.5 mm	≃ 4 to 10 mm	≃ 3 to 7 mm	≃ 12.5 mm
Outline	slightly to markedly curved	falcate	falcate to slightly curved	elongate to ovate, tapering	falcate
Apex	very obtuse	very obtuse	very obtuse to blunt	obtuse	blunt
Base	slightly cordate	emarginate to slightly oblique	slightly cordate to oblique	nearly straight	slightly cordate to emarginate
Midrib: never reaching margin at apex	faint to obsolete	3/5 or more, stout; or faint to obsolete	3/5 or more, stout	4/5 or more, faint to stout	4/5 or more, stout to obsolete
Texture of Aereolae (5)	strong, prominent	strong, prominent	strong, prominent	faint, nondescript	faint; arcuate
Geometry of Aereolae:	4 to 6 sided (?)	4 to 6 sided	4 to 6 sided	4 to 6 sided	4 to 6 sided; arcuate
nearest midrib (5)	≃ 1 to 3 mm long	≃ 2.5 mm long	≃ 1 to 2 mm	≃ 1 to 2 mm	5 to 7 mm long
nearest margin (5)	≃ 0.5 mm long, roundish	≃ 0.25 mm long roundish	≃ 0.2 to 0.6 mm roundish	≃ 0.25 mm long and less	≃ 1 to 2.5 mm long; ≃ 0.5 mm wide
No. of aereolae traversed (5)	4 to 7	8 to 9	5 to 8	4 to 5	9 or 10

(1) Bunbury (1847, p. 427; Pl. 21, Fig. 2A).
(2) Bell (1938, p. 64; Pl. 58, Figs. 4 and 5; p. 64; Pl. 60, Figs. 3 to 5, respectively).
(3) Arnold: 22 complete pinnules in the collection from the Prince and Lingan mines.
(4) Gutbier, *in* Gothan and Weyland (1973, p. 306, Abb. 222).
(5) Measurements according to methods in Fig. 4.

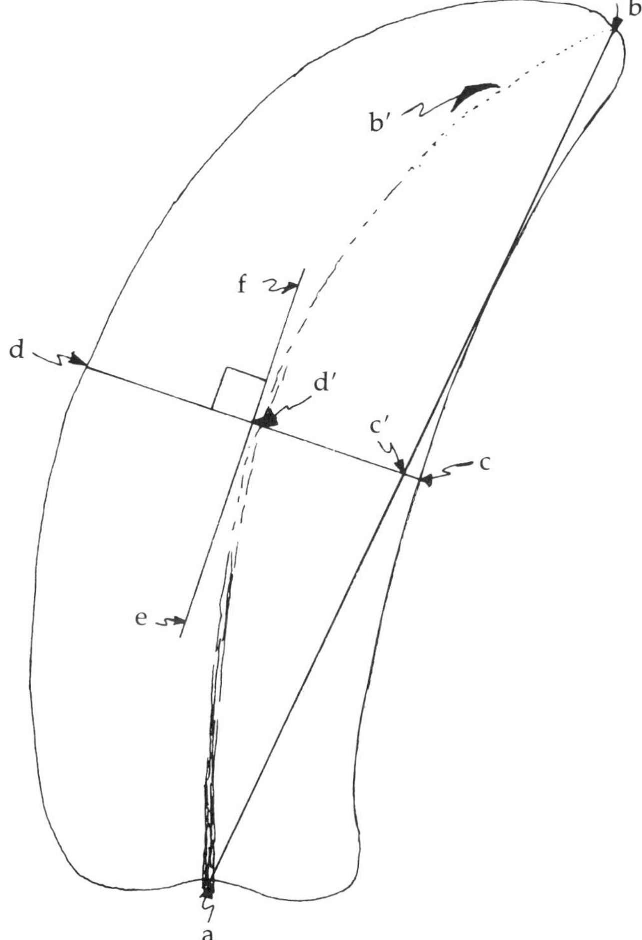

Pinnule of *L. obliqua*

FIGURE 4
Measurements of pinnules of *L. obliqua* which support Table 9. To define the following points: a is on the midrib at the intersection of the midrib and the base; b is on the extended midrib at the intersection of the extended midrib and the apex; if the length (ab)/2 = c′, then the extension of the construction line c′d′ to intersect with the concave margin defines point c, where the line (ef) is tangent to the midrib at d′; and d is found by extending the construction line c′d′ to intersect with the convex margin. The extended midrib or the midrib is designated b′.
Length: mm of line segment ab. If the curvature of b′ is zero, b′ = ab, and ab coincides with ef.
Width: mm of line segment cd. If the curvature of b′ is zero, c′ and d′ coincide.
Aereola nearest midrib: length in mm of one complete aereola not involving the midrib to form a side; longest axis is measured in a linear fashion.
Aereola nearest margin: length in mm of one incomplete aereola which intersects with the concave margin (a linear measurement).
Number of aereolae traversed: the number of aereolae intersecting line cd′ in such a way that an aereola formed by involving a midrib is not counted. The number of aereolae intersecting line cd and not involving aereolae formed by a midrib is generally one more than twice those intersecting cd′.

FIGURE 5
Variation in the branching pattern of vascular strands in *Pecopteris obtusa* Bell, F-514, Emery seam, Glace Bay, N.S. At 1, venation pattern is effectively obscured by the ubiquitous iron oxide in the bootleg pit. At 2, vascular strands fork simply (dichotomy) and ascend strongly to the margin; lobation is absent. At 3, incipient unilateral lobation is evident and the predominant branching pattern in the veins is trichotomous. Position 4, five pinnae closer to base, displays distinct lobation (formation of pinnules) in the lower part of the pinna, and the branching pattern of the laterals in the lobate pinnules is hexad. Pinnae are subopposite with complete confluence at position 1. In the transition from 2 to 4, the laterals, which are strongly ascending at position 2, tend to become perpendicular to the margin at 4. Position 4 represents the lowest pinna of this specimen, and it is interesting to speculate on the nature of the branching pattern of laterals and the nature of pinnule morphology further down the frond. This species is not listed in the catalogue section because of late entry (identification was made in 1977). Refer to Bell (1938, p. 80) who described this species in terms of fructification as *Senftenbergia?*. Dr. Arnold on a more cautious note used pinnule morphology; refer to Table 2.

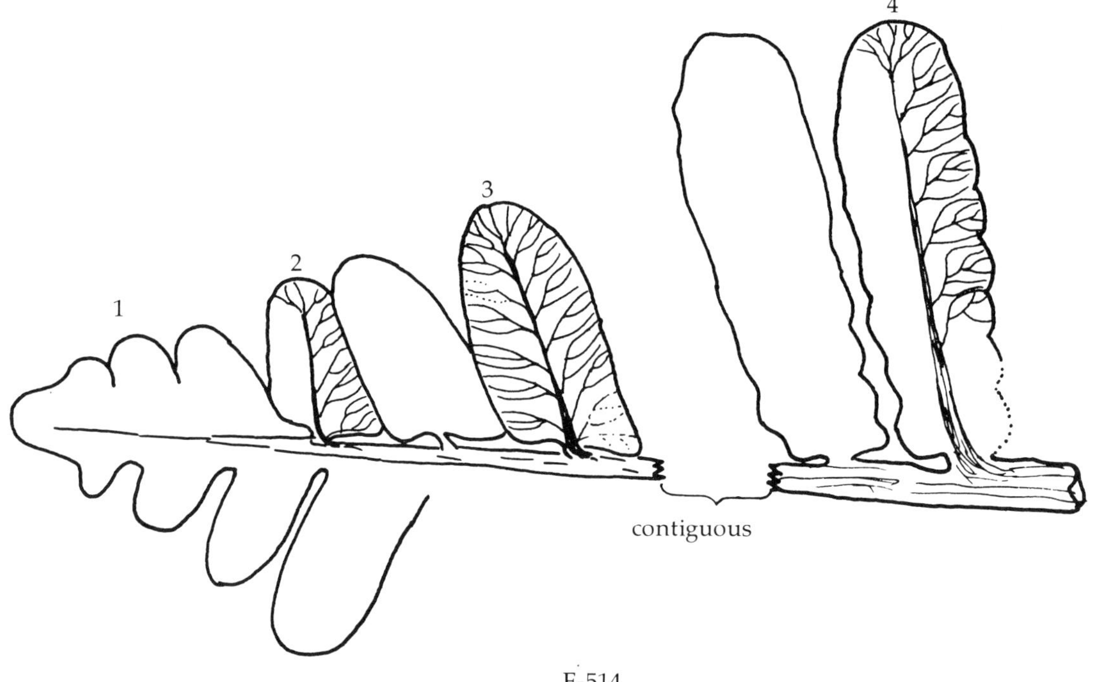

Table 10

Collected fossil plants associated with *L. obliqua* in the Stubbart and Harbour seams of Sydney Coalfield.

	STUBBART SEAM		
Calamarians:			
Asterophyllites equisetiformis	*Calamites carinatus*[1]	*Calamites suckowi*	*Macrostachya infundibuliformis*
Palaeostachya species.			
Lepidodendrids:			
Lepidodendron bretonense	*Lepidodendron pictoense*	*Lepidophloios* species	*Lepidostrobophyllum jenneyi*
Pteridophylls (alethopterids):			
Alethopteris davreuxi	*Alethopteris serli*	*Alethopteris sullivanti*	
(neuropterids):			
Linopteris bunburii	*Neuropteris flexuosa*[2]	*Neuropteris scheuchzeri*	*Neuropteris tenuifolia*
(palaeopterids):			
Eremopteris artemisiaefolia			
(pecopterids):		sphenopterids:	
Odontopteris subcuneata		*Sphenopteris neuropteroides*	
Pteropsids (foliar family Pecopterides):			
Pecopteris (Asterotheca) acadica	*P. (Asterotheca) hemitelioides*	*P. (Senftenbergia) pennaeformis*	*Ptychocarpus unitus*
Sphenopsids:			
Sphenophyllum emarginatum.			
	HARBOUR SEAM		
Calamarians:			
Annularia radiata	*Annularia sphenophylloides*	*Annularia stellata*	*Asterophyllites equisetiformis*
Calamites suckowi			
Lepidodendrids and sigillarians:			
Lepidodendron aculeatum[1]	*Lepidodendron bretonense*	*Lepidodendron dawsoni*	*Lepidodendron pictoense*
Lepidostrobophyllum jenneyi	*Lepidostrobus lanceolatus*[1]	*Sigillaria* cf. *brardi*	*Sigillaria elegans*
Sigillaria species	*Stigmaria ficoides.*		
Pteridophylls (alethopterids):			
Alethopteris serli	*Alethopteris valida*		
(neuropterids):			
Cyclopteris cf. *fimbriata*	*Neuropteris (Mixoneura) flexuosa*	*Neuropteris flexuosa*[2]	*Neuropteris macrophylla*
Neuropteris rarinervis	*Neuropteris scheuchzeri*[3]	*Neuropteris (Mixoneura) ovata*	
(pecopterids):			
Mariopteris latifolia	*Mariopteris nervosa*	*Odontopteris minor*	*Odontopteris subcuneata*
Pecopteris obtusa			
(sphenopterids):			
Sphenopteris neuropteroides	*Sphenopteris (Diplotmema) whitii*		
Pteropsids (foliar families Pecopterides and Sphenopterides):			
Asterotheca cf. *abbreviata*[1]	*Asterotheca daubreei*	*Asterotheca herdi*	*Asterotheca oreopteridia*[1]
Pecopteris (Asterotheca) acadica	*P. (Senftenbergia) penaeformis*	*Ptychocarpus unitus*	
Sphenopteris (Oligocarpia?) crenatodentata			
Sphenopsids:			
Sphenophyllum cuneifolium	*Sphenophyllum emarginatum*	*Sphenophyllum majus*	*Sphenophyllum oblongifolium.*

1 Not previously reported from the Sydney Coalfield.
2 Forma *magna*.
3 Var. *nordfrancia* and var. *dawsoni*, neither of which have previously been reported from the Sydney Coalfield.

Neuropteris (Mixoneura) flexuosa Sternberg (F-5, F-255, F-266-68, -70, F-274, F-281, F-283).

Plate 22, Fig. 2; Plates 23, 24

FOSSIL ASSOCIATIONS:
F-266-68, 270 (on same block): with a calamarian cone, probably *Palaeostachya*, F-269, F-271; *Alethopteris davreuxi* (Brongniart); F-272; *L. obliqua* (Bunbury), and *Sphenophyllum* sp. indet.

SAMPLE LOCATIONS
F-5; Unknown coal measure in the Morien series, Cape Breton.
F-255, F-266-68, 270 (on one block), F-274, F-281; F-283: Stubbart seam, Point Aconi, Prince Mine.

BELL'S RANGE: L-Pt-zones.

REMARKS:
Note association of this species with *Alethopteris davreuxi* and *L. obliqua* (in Pt-zone) on block F-255. Classification is according to Bell (1938, p. 55). Other workers, i.e., Crookall, 1959, v. 4, Pt. 2, p. 149-162, place it in *N. ovata* forma *flexuosa*. According to Bell, F-270 is *N. flexuosa* forma *magna;* in particular refer to *N. flexuosa* Sternberg forma *magna*. The subgeneric taxon *(Mixoneura)* is a modification of *Neuropteris* and not a distinct morphological type (Arnold, c.f. footnote Table 1). Weiss was never given priority for the sub-genus *Mixoneura* Weiss, 1869.

Today's definition of *(Mixoneura)* or the mixoneurid condition differs little from P. Bertrand's re-interpretation:

a) in this form of *Neuropteris* (form-genus) several pairs of pinnules immediately below the terminal pinnule on a pinna are broadly attached (with the remaining pinnules on the pinna of cordate or subcordate attachment), and

b) the lowermost (basal) veins of the broadly attached pinnules arise from the rachis,

c) thus resembling *Odontopteris*.

If this condition extends only to one pair of pinnules (rather than several) we may speak of the mixoneuroid condition; however, this may be pedantic (and does not apply to Havlena's quote below). Arnold (1949, p. 196) reported six pairs of pinnules of mixoneurid condition on one pinna of *N. obliqua*. This species is of the impar form (pinnules are alternately borne on the rachis and *one* pinnule terminates the pinna). See *N. heterophylla*,

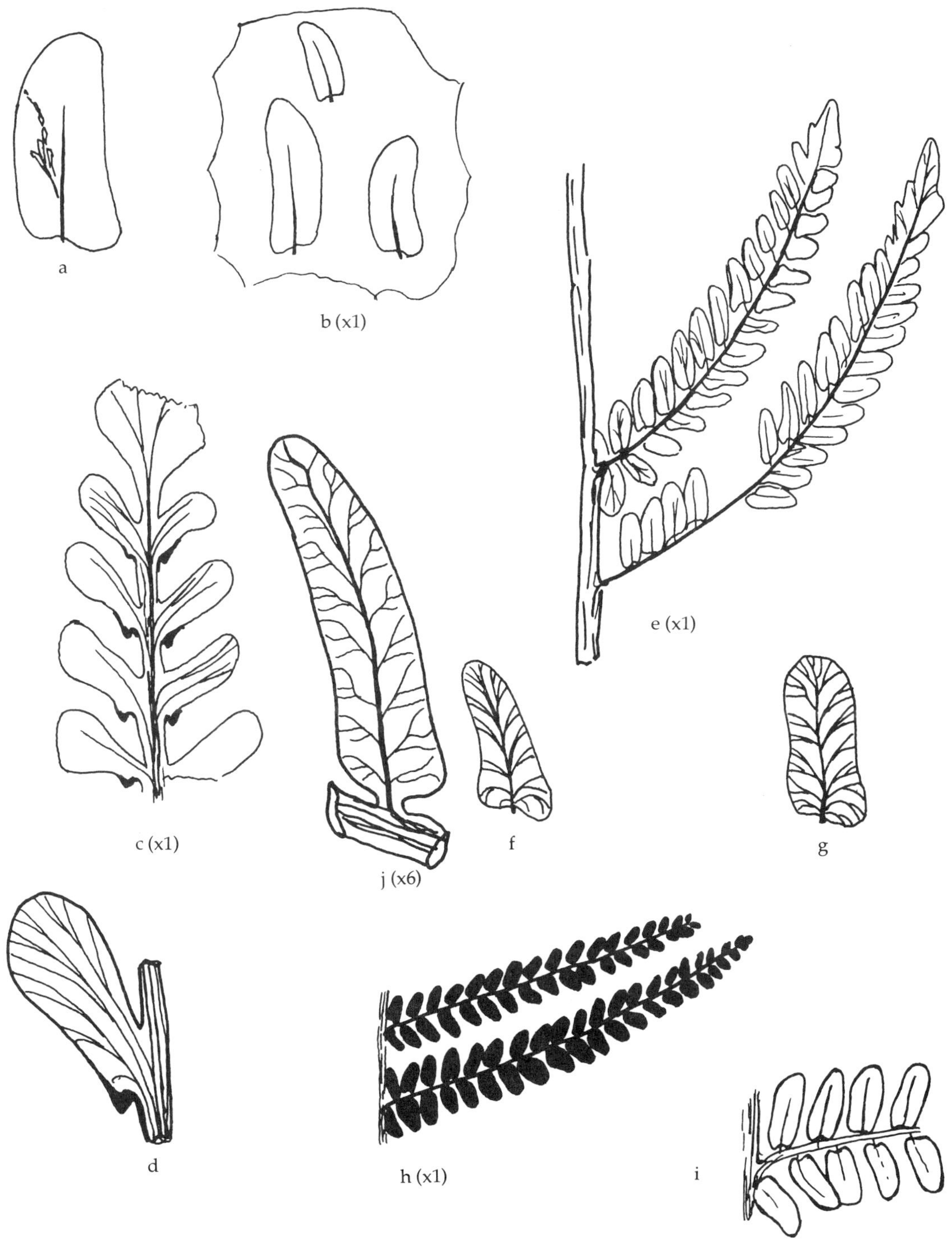

FIGURE 6
Redrawn fern-like foliage from Bunbury, Lindley and Hutton, and present collection. a: enlarged pinnule of Bunbury's *Linopteris obliqua* showing venation schema partly redrawn. b: outline of pinnules of Bunbury's *L. obliqua* (note the obtuse apices). c and d: *Odontopteris subcuneata* and enlarged detail, respectively, of one pinnule of Bunbury's species. e: partly redrawn frond of Bunbury's *Neuropteris rarinervis,* with f and g showing details of vein structure. h: *Neuropteris attenuata,* a partly redrawn frond. i: detail of the same of Lindley and Hutton's species. j: *Neuropteris rarinervis* (see Fig. 1).

Sources: a-g Bunbury (1847, Pl. XXI, Figs. 2A, 2B. Pl. XXIII, Figs. 1A, 1B. Pl. XXII, Figs. 1A, 1B). h and i: Lindley and Hutton (1837, Pl. 174).
j: present collection

N. ovata and *N. tenuifolia* for examples of the mixoneurid condition. However, Havlena (1953, p. 164) stated "In the Bohemian regions of the Permocarboniferous I have established ten typical (not mixoneuroid) neuropterid species: . . . *N. heterophylla* Brongniart, *N. schlehani* Stur, *N. tenuifolia* (Schlotheim) . . .".

▶ *Trigonocarpus* sp. indet. (attached seed).

Recently collected specimens from the Lingan Mine, Harbour seam (F-737) show that the seed of the *Trigonocarpus* type is attached to a pinna rachis. The foliage to which the seed is attached is neuropterid and identified as *Neuropteris (Mixoneura) flexuosa* (the block contains other neuropterid foliage). Refer to Stewart and Delevoryas (1956) for detailed discussion on attachments of seeds in medullosan pteridosperms.

Therefore, we propose a new classification for the species *N. (Mixoneura) flexuosa* putting it in Table 4 of the seed ferns, and thus removing it from the pteridophylls in Table 1, (Zodrow and McCandlish, 1979a).

Neuropteris flexuosa Sternberg forma *magna* (F-270, F-375, F-450, F-459, F-508).

Plate 25

FOSSIL ASSOCIATIONS:
F-270: with *N. (Mixoneura) flexuosa* Sternberg.
F-450; with *N. scheuchzeri* Hoffmann.

SAMPLE LOCATIONS:
F-450, F-459, F-508: Emery seam.
F-375: Harbour seam, Lingan Mine dump.
F-270: Stubbart seam, Prince Mine.

REMARKS:
F-508 is a specimen of a terminal pinna in which the mixoneuroid condition is traceable at most to the first pair of the impar form pinnules; more likely, it is *N. (Mixoneura) ovata*. The specimen lacks diagnostic detail. F-375 is *N.* cf. *gigantea*, a detached pinnule (a pinna with terminal parts is required for definite determination).

Neuropteris heterophylla (Brongniart) (F-68, F-445, F-446).

Plates 26, 27; Plate 28, Fig. 1

SAMPLE LOCATIONS:
F-445, F-446: Emery seam.
F-68: Mc Aulay seam.

BELL'S RANGE: L-Pt-zones.

REMARKS:
These two terminal fronds (F-445, F-446) of 8 to 15 cm respectively, are in good condition. The mixoneurid condition is evident over, at most, two pairs of pinnules in these specimens.

This species is of the impar form. Complete fronds with attached cyclopterid pinnules are reported from the Carboniferous in the Ruhrgebiet of Germany (Gothan and Weyland, 1964).

Brongniart (1830, p. 243) described the species thus:

N. foliis maximis tripinnatis, quandoque e basi bifurcatis, pinnis alternis magis minusve elongatis, superioribus brevissimis; pinnulis forma diversissimis, pinnarum inferiorum oblongis sublobatis, intermediarum ovatis, superiorum subrotundis minimis paucioribus; terminalibus oblongo-lanceolatis, basi cuneatis, lateralibus multo longioribus; omnibus basi cordatis, nervulis arcuatis tenuissimis.

We have reasons to believe that this species is also present in the basal part of the Morien series but it is difficult to be sure from the isolated pinnules (F-68, F-78) collected from the Mc Aulay seam. See *N. tenuifolia* for remarks on the relationship between the two species.

Neuropteris macrophylla Brongniart (F-637, F-642-1).

Plate 43, Fig. 2

SAMPLE LOCATION:
Lingan Mine dump, Harbour seam.

BELL'S RANGE: Pt-zone.

REMARKS:
F-637 is a complete frond of about 20 cm, in good condition. Bell remarked that his specimen came from the highest strata in the Sydney Coalfield. This species may be confused with *Neuropteris scheuchzeri*. It differs from the latter in the absence of adpressed

hair structure, the absence of symmetrical small leaves near the rachis, and it is smaller than form *angustifolia.*

The terminal pinnule of this frond is about 2 cm long, ovate in shape and followed directly by a pair of pinnules showing mixoneuroid conditions.

▶ *Neuropteris (Mixoneura) obliqua* (Brongniart) [Stockmans, 1933, Pl. 8, Fig. 6] (F-474).

Plate 28, Fig. 2

FOSSIL ASSOCIATIONS:
With *N. (Mixoneura) ovata* Hoffmann, and *Lepidodendron bretonense* Bell.

SAMPLE LOCATION:
Phalen seam.

REMARKS:
This is a 5.4 cm specimen of the terminal portion of a pinna; the identification is questionable. Not previously reported from Sydney Coalfield.

Neuropteris (Mixoneura) ovata Hoffmann (F-20, F-30, F-32, F-126, F-127, F-129, F-132, F-188, F-379, F-444, F-640, 967G10.66).

Plate 28, Fig. 3, Plates 29-32; Plate 33, Fig. 1

FOSSIL ASSOCIATIONS:
F-129: with *Neuropteris scheuchzeri* Hoffmann.
F-379: with *Linopteris obliqua* (Bunbury).

SAMPLE LOCATIONS:
F-126, F-127, F-129, F-132, F-444: Emery seam.
F-188: Phalen seam.
F-20, F-30: Harbour seam, #12 Mine dump.
F-379, F-640: Harbour seam, Lingan Mine dump.
F-32, 10.66: Unknown locations in the Morien series, Cape Breton Island.

REMARKS:
This species is abundant at the Emery seam, Glace Bay and at the Harbour seam, Lingan, N.S. All listed specimens have apical parts, except 967G10.66 which is a branching stalk, striated longitudinally, 2 cm diameter and 15 cm long, and F-379 which is a 13 cm frond of well-preserved quality. The mixoneurid condition extends over more or less four pairs of the impar pinnate specimens. See *Cyclopteris fimbriata* for association and time range.

Bell (1938, Plate 55, Fig. 1) equated this species with *C. fimbriata.*

Neuropteris cf. *pseudogigantea* H. Potonié (F-69 to F-75).

Plate 33, Fig. 2

SAMPLE LOCATION:
Mc Aulay seam**.

BELL'S RANGE: Lch?-L-zones.

REMARKS:
Doubts about the specific identity are based on the dimension of the readily caducous pinnules: approximately 1.5 cm long and 5-7 mm wide; Bell (1938, p. 56) lists 1.7-1.9 cm long and 7 to 9.5 mm wide, respectively.
Neuropteris aculeata Bell is a synonym of *N. pseudogigantea* (Bell, 1962, p. 39).

**Spelling of Mc Aulay *(sic)* may be incorrect on the Geology Map 362A, Glace Bay Sheet, 1938. However, we do not wish to alter a previously published name at this point. See also Table 6.

Neuropteris rarinervis Bunbury (F-86, F-363-1, F-374, 967G20.4, 967G36.1, F-567).

Plates 34-36; Plate 37, Fig. 1

SAMPLE LOCATIONS:
F-86: Shoemaker seam.
F-567: Emery seam.
F-363-1, F-374: Harbour seam, Lingan Mine dump.
20.4: two miles east of Stellarton, McLellan Brook, Pictou series.
36.1: Morien series, Cape Breton.

BELL'S RANGE: L-Pt-zones.

REMARKS:
One specimen is a frond, all others are detached pinnae.
The identity of F-86 is uncertain; see Fig. 1 for F-567. The specimen 36.1 is a 15 cm upper part of a frond. This species is of the impar form. Bunbury was the first to describe and figure it in North America (Bunbury, 1847). The specimens originated from Cape Breton Island.

Bunbury described his species thus:

Neuropteris rarinervis (n. sp.)

Spec. Char. N. fronde bipinnata, pinnulis contiguis obliquis oblongis apice rotundatis, basi oblique subcordatis subauriculatis, margine subundulatis; terminali majore hastatodeltoidea subtrilobata; venis remotis arcuatis bis furcatis. (Bunbury, 1847, p. 425).

Lindley and Hutton (1835, v. 3, Plate 174) wrote:

Neuropteris attenuata

A coal measure fern, allied to *Neuropteris Loshii,* from which and all other species it differs in its leaflets becoming gradually smaller, till the terminal one is less than any of the others. This is an unusual circumstance in the genus *Neuropteris* and distinctly marks the species.

Stockmans (1933) proposed in his Conclusion that the name *N. attenuata* should replace *N. rarinervis. N. attenuata* merits priority by virtue of earliest publication, at least for the specimens from Illinois; Stockmans did not see Bunbury's specimens (Stockmans, 1933, p. 20-23, 51). Bunbury (p. 426) was aware of *N. attenuata* and stated that it seems to be the nearest to his species. Although the venation in *N. attenuata* appears to resemble that of *N. rarinervis,* there is sufficient difference in the terminal pinnule of *N. rarinervis,* which is larger and halberd-shaped (as a consequence of the mixoneurid condition?), to warrant specific status (see Fig. 6).

Havlena (1953, p. 135) for the first time cited clearly distinguishing characteristics between the two species:

	N. attenuata	*N. rarinervis*
Pinnules	alternating, ovate-triangular to ovate (convergent to the tip)	alternating, elongated elliptic to tongue-shaped elliptic (subparallel margins); at the base on the catadrome side more developed
Pinnule outline	triangular	tongue-shaped
Pinnule tip	rounded, mostly gently to considerably sickle-shaped; arched towards the tip of the pinna	broadly rounded
Pinnule base	very moderately cordate	cordate, with the auricula on the catadrome side
Pinnule attachment	with a contracted base, . . ., apparently with the whole base	with a contracted base
Pinnule length	4 to 8 mm or more	5 to 12 mm
Pinnule width	3 to 5 mm or more	5 to 6 mm
Terminal leaflet	pointed	bluntly pointed
Attachment of the pinnules at the top of the pinnae	odontopteroid (to alethopteroid)	odontopteroid
Remarks		the leaflets have a rigid appearance (Bunbury, 1847, p. 426; Bell, 1938, Pl. 53, Fig. 2).

The reader is invited to apply Havlena's criteria to our figured specimens, and to certain parts of Fig. 6 in this text.

Neuropteris scheuchzeri Hoffmann (F-13, F-133, F-136, F-315, F-316, F-317, F-318, F-387, F-408, F-410, F-449, F-452, F-453, F-468, F-501, F-564).

Plate 37, Figs. 2, 3; Plates 38-42

FOSSIL ASSOCIATIONS:
F-318: with *Odontopteris* sp. indet., F-318-1.
F-410: with *Lepidophyllum* sp. indet., F-411 (verso of F-410).
This species also occurs together with:
F-150: *Alethopteris serli* (Brongniart), Emery seam.
F-325: *Sphenopteris (Diplotmema) whitii* Bell, Emery seam.
F-326: crushed seed of an alethopterid, Harbour seam.
F-427: *Asterotheca daubreei* Zeiller, Harbour seam.
F-162: *Pecopteris (Asterotheca) acadica* Bell, Emery seam.
F-390: *Lonchopteris eschweileriana* Andrae, Emery seam.

SAMPLE LOCATIONS:
F-133, F-136, F-387, F-449, F-452, F-453, F-564: Emery seam.
F-468: Phalen seam.
F-315, F-316, F-317, F-318: Harbour seam, Lingan Mine dump and #12 Mine dump.
F-501: Sydney Mines, Harbour seam.
F-408, F-410: Upper (?) Bonar seam, Point Aconi.
F-13: Unknown coal measure in the Morien series, Cape Breton Island.

BELL'S RANGE: Morien series.

REMARKS:
F-501 is on permanent loan to the College from the Beaton Institute, College of Cape Breton, Sydney, N.S.

In Sydney Coalfield, this species is a ubiquitous fossil, the variation of which is documented in accompanying plates. Bunbury (1847, p. 423-24) quite rightly used the name "heterophylla" for this species. The modifier forma *angustifolia* is also a good description for some specimens (F-453, for example). In addition, this species is associated with *N. (Mixoneura) ovata*, *N. heterophylla* and *Alethopteris valida* in the roof of the Emery seam. See block F-483 for its association with *Asterophyllites equisetiformis*.

All specimens are single detached pinnules, except for F-387, which is attached to a parallel-striated rachis of 3 mm displaying densely distributed small punctae whose origin may be related to bases of hair-type structures, and F-501, which are assumed to have been attached pinnules. This represents a simple-pinnate frond 28 cm long.

F-452 is a dichotomous leaf of 3.5 cm length. The midvein divides at about 2 cm from the cordate base, giving rise to two separately-pointed lobes of unequal length on the base of the pinnule. The venation is neuropterid. This variation has not been previously reported from Sydney Coalfield.

In specimen F-453, 5 cm long, two symmetrically paired ovate leaves of 8 mm are situated about the cordate base. The ovate leaves are clearly separated from the main leaf by 3 mm on the extension of the midvein of the leaf towards the rachis. Bunbury called it *Neuropteris cordata* var. *angustifolia* (Bunbury, 1847, Pl. XXI, Fig. 1B). Refer to F-430 for association with cyclopterid species.

This species is named after choirmaster and doctor of medicine Herr J. J. Scheuchzer of Zurich. Doctor Scheuchzer wrote the earliest serious work on paleobotany, *Herbarium diluvianum,* in 1709 (Wendt, 1965). Another species named after Dr. Scheuchzer is a cinnamon tree of the early Tertiary, *Cinnamomum scheuchzeri* Heer (Fraas, 1976, p. 205, Pl. 60, Fig. 6).

From the Emery seam at Steele's Road we have collected a specimen that may be designated as *N. scheuchzeri* var. *nordfrancia* P. Bertrand; see Havlena's Plate 8, Fig. 4 (1953).

▶ This variety has not been previously reported from Sydney Coalfield.

▶ *Neuropteris scheuchzeri* var. cf. *dawsoni* (F-336).

Plate 43, Fig. 1

FOSSIL ASSOCIATIONS:
With *Linopteris muensteri* (Eichwald), F-334, and *Alethopteris davreuxi* (Brongniart), F-335.

SAMPLE LOCATION:
Harbour seam at Lingan Mine dump.

REMARKS:
One single detached pinnule of about 5 cm. Not previously recorded from Sydney Coalfield.

▶ *Neuropteris* cf. *schlehani* Stur (976GF26.1, F-472).

Plate 44, Figs. 1, 3

SAMPLE LOCATIONS:
26.1: unknown.
F-472: Phalen seam.

REMARKS:
Specimen 26.1 is a partial frond of 6 cm, probably a small form of the species; F-472 is the center part of a pinna, and specific identification is uncertain without its terminal parts. Not previously recorded from Sydney Coalfield.

Neuropteris sp. indet. (F-85, F-484, 967G20.3).

Plate 45, Fig. 1

SAMPLE LOCATIONS:
F-85; Shoemaker seam.
F-484: Harbour seam, Lingan Mine dump.
20.3: McLellan Brook, Pictou series.

REMARKS:
F-484 shows characteristics of several species; it could be *Neuropteris obliqua*.

Neuropteris tenuifolia (Schlotheim) (F-185-1, F-465, F-480, F-614, F-630, F-631).

Plate 44, Fig. 2; Plate 46

FOSSIL ASSOCIATIONS:
F-185-1: with *N. (Mixoneura) flexuosa* Sternberg or *N. (Mixoneura) ovata* Hoffmann. See also block F-483, *Asterophyllites equisetiformis* (Sternberg).

SAMPLE LOCATIONS:
F-465; Shoemaker seam.
F-630, F-631: Tracy seam.
F-185-1, F-480: Phalen seam.
F-614: Stubbart seam, Prince Mine.

BELL'S RANGE: Lch-L-zones.

REMARKS:
F-185-1, an 8 cm pinna, is split longitudinally. The mixoneuroid form is observable in the specimens. This species belongs to the impar form and is relatively abundant at the roof of the Phalen seam where it is found in grey micaceous sandstone.

This species is highly variable and may resemble *N. heterophylla* and *N. flexuosa;* see specimens figured by Stockmans (Stockmans, 1933, Pl. 2, 3, and 4). Some authorities question the distinction between this species and *N. heterophylla,* believing that the two species represent different parts from one "large, variable" frond (Arnold, 1949, p. 194-5).

Stockmans (1933) claimed the two species were distinct but left doubt about the distinguishing characteristics (fragments alone cannot be satisfactorily assigned). However, Havlena (1953, p. 140) asserted that Stockmans (1933) had confirmed independence of *N. tenuifolia* and *N. heterophylla*.

Bell (1938, p. 54) believed that specimens from Sydney Coalfield were slightly different from Schlotheim's original description of this species; in particular, he believed that these specimens resembled more closely the European variety *nordfrancia*. However, Professor Arnold left no doubt that specimens in the collection are the same species that Schlotheim described (*in litt.*, May, 1977).

PALAEOPTERIDES

Adiantites adiantoides Lindley and Hutton (967G123.26).

SAMPLE LOCATION:
Gaspereau, Kings Co., N.S. (Horton series).

REMARKS:
According to Professor Arnold (c.f. footnote 4), this is perhaps the most artificial of the five families; it may have existed during Devonian time with extension into and through the Pennsylvanian Period. Pinnules are round, ovate, rhomboidal or wedge-shaped and often arranged in even rows on the rachis. Lack of a conspicuous midrib seems diagnostic in conjunction with one or more vascular strands entering the pinnule through its constricted base. These laterals divide in a fan-like manner and traverse the pinnule with little or no curvature.

Eremopteris artemisiaefolia (Sternberg) [Kidston, v. 2, Pt. 5, p. 407].
(F-105, F-288 verso, 967G10.4, F-622, F-639, F-643).

Plate 45, Figs. 2, 3; Plate 47, Figs. 2-4

FOSSIL ASSOCIATION:
F-288: with *Linopteris obliqua* (Bunbury), L.O.-19, -20.

SAMPLE LOCATIONS:
F-105: Tracy seam.
F-643: Emery seam.
F-288 verso, F-639; Harbour seam, Lingan Mine dump.
F-622: Stubbart seam, Prince Mine.
10.4: Morien series, Cape Breton Island.

BELL'S RANGE: L-zone (in the roof of Tracy seam).

REMARKS:
F-105 is a terminal pinna of 2 cm (mounted specimen); F-288 is also a terminal pinna (3 cm). Both specimens are well preserved. 10.4 is a 6 cm part of a frond.

D. White regarded this species as Lower Pennsylvanian (White, 1943, p. 92).

PECOPTERIDES

Callipteridium sullivanti (Lesquereux) (F-275, 967G16.2, 3, 7, 8, 10, 11).

Plate 53, Figs. 4, 5; Plate 55, Figs. 2, 3

FOSSIL ASSOCIATIONS:
F-275: with *L. obliqua* (Bunbury) and *Sphenophyllum* sp. indet.

SAMPLE LOCATIONS:
F-275: Stubbart seam at Point Aconi, Prince Mine.
16.2, 3, etc.: Harbour seam, Florence Mine.

BELL'S RANGE: L-zone (Mullins seam, Table 6).

REMARKS:
F-275 is a 6 cm pinna with terminal parts missing. As with all specimens from the Prince Mine, this is also well preserved.

The genus *Pecopteridium*, as in *P. sullivanti*, is preoccupied, hence *Callipteridium sullivanti*. Gothan and Weyland (1973, p. 289) cited the following synonymy:

C. sullivanti = *Callipteris sullivanti* = *Alethopteris sullivanti* Schimper

Mariopteris latifolia (Brongniart) (F-494).

Plate 48; Plate 49, Fig. 1

FOSSIL ASSOCIATIONS:
With *Ptychocarpus unitus* (Brongniart) and *Pecopteris (Senftenbergia) pennaeformis* Brongniart, F-495.

SAMPLE LOCATION:
Harbour seam, Lingan Mine dump.

BELL'S RANGE: L-Pt-zones.

Mariopteris nervosa (Brongniart) (F-42, F-158, F-194, F-291a, F-344, F-368, F-435, F-438, F-469, F-470).

Plate 49, Figs. 2, 3; Plate 50

FOSSIL ASSOCIATION:
F-344: with *L. obliqua* (Bunbury).

SAMPLE LOCATIONS:
F-158: Emery seam at Glace Bay.
F-194, F-469, F-470: Phalen seam.
F-291a, F-344, F-368, F-435, F-438: Harbour seam at Lingan Mine.
F-42: Harbour seam, #12 Mine dump, New Waterford, N.S.

BELL'S RANGE: L-Pt-zones.

REMARKS:
F-435 is an ultimate frond (5 cm); pinnules are completely confluent and pinnae closest to penultimate resemble single pinnules at lower parts of the frond e.g. F-194 and F-469, and F-470. The remaining specimens are mainly fragments of fronds.

Bell (1962, p. 26, and 29) specifically differentiates and divides *M. nervosa* Brongniart into *M. carnosa* Corsin and *M. hirta* (Stur).

On the basis of immersion of veins, Bell (1962) assigned most material from Nova Scotia to *M. carnosa* Corsin. To specifically separate the present material, Bell's plesiotypes are necessary for study.

Mariopteris tenuis Bell (967G10.16).

Plate 51

FOSSIL ASSOCIATIONS:
With *Annularia stellata* (Schlotheim) and with *Neuropteris (Mixoneura)* cf. *ovata* Hoffmann.

SAMPLE LOCATION:
Morien series, Cape Breton Island, N.S.

REMARKS:
A 5 cm partial frond split longitudinally, with ultimate parts missing.

Odontopteris minor Brongniart (967G10.25, F-503).

Plate 47, Fig. 1; Plate 52

SAMPLE LOCATIONS:
10.25: Mabou Coal Mines, Mabou, Cape Breton Island, N.S.
F-503: Phalen seam.

BELL'S RANGE: L-Pt-zones.

REMARKS:
This species is specifically differentiated from *O. subcuneata* by triangular pinnules especially well developed in the terminal parts of fronds. Fig. 3 of Pl. 52 shows a good triangular pinnule in the center of that specimen. Compare to Bell's specimens, especially that of Pl. 56, Fig. 5 (Bell, 1938, p. 61-62). According to Gothan and Weyland (1973, p. 290) this species is of late Pennsylvanian and early Permian age in Europe.

Odontopteris subcuneata Bunbury (F-87-1, 2, F-120, F-123, F-124, F-273, F-318-1, F-418, F-580, 967G10.53).

Plate 53, Figs. 1-3; Plate 54; Plate 55, Fig. 1

FOSSIL ASSOCIATIONS:
F-318-1: with *Neuropteris scheuchzeri* Hoffmann, F-318.
F-120: with *Triletes auritus* var. *grandis* Zerndt, F-120-1.

SAMPLE LOCATIONS:
F-87-1, 2: Shoemaker seam.
F-120, F-123, F-124, F-580: Emery seam.
F-318-1: Harbour seam at Lingan Mine dump.
F-273, F-418: Stubbart seam, Point Aconi.
10.53: Morien series, Cape Breton Island.

BELL'S RANGE: L-Pt-zones.

REMARKS:
The identity of F-87-1 is not certain.

This species is rare at the Emery seam, but apparently abundant at the Stubbart seam outcrops west of Point Aconi on Boularderie Island, Fig. 12.

Bunbury's description of this species is as follows:

Odontopteris subcuneata (n. sp.)

Spec. Char. O. pinnulis remotis subopposititis decurrentibus oblique obovato-cuneatis subrecurvis apice rotundatis: terminali majore ovata; rachi lata marginata; venis tenuissimis arcuatis dichotomis. (Bunbury, 1847, p. 427).

Pecopteris sp. indet. (F-193).

Plate 56, Fig. 1

SAMPLE LOCATION:
Phalen seam.

SPHENOPTERIDES

Sphenopteris neuropteroides (Boulay) (F-22, F-156: see verso F-157, F-371, F-372, F-373, F-428, F-429, F-441, F-534, F-543, F-544, F-573, F-659, 967G143.11).

Plate 56, Figs. 2-4; Plates 57, 58

FOSSIL ASSOCIATION:
F-429: with *Annularia sphenophylloides* (Zenker), F-442.

SAMPLE LOCATIONS:
143.11: Old Gowrie Pit near Port Morien.
F-156: See verso F-157, F-534, F-543, F-544, F-573: Emery seam, Glace Bay.
F-22: Harbour seam at #12 Mine dump.
F-371, F-372, F-373, F-428, F-429, F-441: Harbour seam, Lingan Mine dump.
F-659: Stubbart seam, Prince Mine.

BELL'S RANGE: L-Pt-zones.

REMARKS:
F-22 is a terminal pinna with well preserved venation; the identity is questionable. F-659 is a beautiful 17 cm frond. The identities of apical fronds F-543 and F-544 are questionable.

This family is distinguished from the palaeopterids by pinnule characteristics. In the sphenopterids there is commonly a central vein in the lower part of the pinnule from which the laterals flow. The laterals then, with noticeable curvature, pass to the margin. The pinnules are generally broadly attached (Fig. 3 parts h and i).

Sphenopteris sp. indet. (F-404, F-405, F-422).

Plate 59, Fig. 1

SAMPLE LOCATIONS:
F-404, F-405: Harbour seam, #26 Mine dump.
F-422: Harbour seam, underground Lingan Mine.

REMARKS:
F-404, F-405 are imprints of large (8 cm long, 2 cm across) frond stalks and flattened petioles which bear distinct, large punctae.

Sphenopteris cf. *striata* Gothan (F-98-99, F-101, F-107, -109).

Plate 59, Figs. 2, 3; Plate 60

FOSSIL ASSOCIATION:
F-98-99: with *Lepidodendron dawsoni* Bell, F-97b.

SAMPLE LOCATIONS:
F-107, -109: Tracy seam.
F-98-99: F-101: Shoemaker seam.

BELL'S RANGE: Morien series.

REMARKS:
Striations on F-107-109 do not show, but fan-shaped pinnules agree with those shown by Bell (Bell, 1938, p. 22). F-101 is a 5 cm frond section with two attached pinnae. The preservation of F-98-99 and F-101 is not good, but the striations can be seen with the naked eye. This species is abundant at the Shoemaker seam.

Sphenopteris cf. *suspecta* D. White (F-412).

Plate 61, Fig. 1

SAMPLE LOCATION:
Upper? Bonar seam, Point Aconi.

BELL'S RANGE: L-zone.

Sphenopteris (Diplotmema) whitii Bell (F-102, F-321, F-325, F-423).

Plate 61, Figs. 2, 3; Plate 62, Fig. 1

FOSSIL ASSOCIATIONS:
F-325; with *Neuropteris scheuchzeri* Hoffmann, and a crushed alethopterid seed case, both on block F-326.

SAMPLE LOCATIONS:
F-102: Shoemaker seam.
F-321, F-325: Harbour seam at Lingan Mine dump.
F-423: Harbour seam, Lingan Mine, underground.

BELL'S RANGE: L-Pt-zones.

REMARKS:
With the exception of F-423, all specimens are partial fronds. This genus was defined by Stur in 1877 (p. 130) as having a frond structure in which the main rachis forks at a wide angle below the place of attachment of the lowest secondary pinnae, implying bilateral symmetry in the frond (plane of symmetry).

FILICINEAE

PECOPTERIDES

▶ *Asterotheca* cf. *abbreviata* (Brongniart) [Crookhall R., 1929, Fig. D. Pl. XXVI] (F-1, F-421).

Plate 62, Fig. 2; Plate 63

SAMPLE LOCATIONS:
F-1: Unknown coal measure of the Morien series, Cape Breton Island.
F-421: Harbour seam at Lingan Mine (underground).

REMARKS:
F-1 is a 22 cm frond with ultimate parts missing; it appears quite different from Bell's *A. miltoni* forma *abbreviata* and also shows some resemblance to *Acitheca polymorpha*.

F-421 is a fertile pecopterid frond (ultimate parts missing) of 15 cm; it resembles *Asterotheca abbreviata* but specific identity is not certain. Not previously recorded from the Sydney Coalfield.

▶ *Asterotheca daubreei* Zeiller [Kidston, v. 2, Pt. 6, Pl. CXXV] (F-427, F-431, F-432).

Plates 64, 65; Plate 66, Fig. 1; Plate 67, Fig. 1

FOSSIL ASSOCIATIONS:
F-427; with *L. obliqua* (Bunbury) and *N. scheuchzeri* Hoffmann.

SAMPLE LOCATION:
F-427, F-431, F-432: Harbour seam at Lingan Mine.

REMARKS:
F-427 is a 27 cm fertile frond. F-431 is detached pinnae some of which are vegetative, others fertile. F-432 is a beautiful 11 cm frond assumed to be vegetative; note its similarity to *Asterotheca abbreviata*. All specimens are in good condition.

A re-evaluation of F-427 indicates that it may be *A. herdi.* Of the three distinctions cited by Bell (Bell, 1938, p. 73) between *A. daubreei* and his *A. herdi* sp. one is absent in F-427: punctae (*in litt.*, Dr. Arnold, Oct., 1975); however, see Plate 65, Fig. 2. Similar comments apply to F-431 and F-432. A newly discovered species at the Sydney Coalfield.

Asterotheca herdi Bell (F-287-1, -2, F-413, F-414, F-486).

Plate 66, Figs. 2, 3; Plate 67, Fig. 2; Plates 68, 69

SAMPLE LOCATIONS:
F-486: Harbour seam, Lingan Mine dump.
F-287-1, -2: Stubbart seam, Point Aconi, Prince Mine.
F-413, F-414: Bonar seam, Point Aconi.

BELL'S RANGE: L-Pt-zones.

REMARKS:
F-413 is a single detached 8 cm pinna; specimen F-414 represents two fragmental fronds of about 6 cm each, and F-486 is a well-preserved partial frond. Refer to *A. daubreei* for differentiation between the two species.

This species is named after Walter Herd, mining engineer with the former Dominion Steel and Coal Corporation in Sydney, N.S.

▶ *Asterotheca oreopteridia* Schlotheim [Kidston, v. 2, Pt. 5, p. 495-500] (F-189, F-314, F-327, F-392).

Plate 70; Plate 71, Figs. 1, 2

SAMPLE LOCATIONS:
F-189: Phalen seam.
F-314, F-327: Harbour seam at Lingan Mine.
F-392: Harbour seam at #26 Mine dump.

REMARKS:
F-189 is a 6 cm partial frond with ultimate parts missing. F-314 is a 15 cm frond; F-327 seems to be the tip of a frond, 7 cm; while F-392 is a 4 cm partial pinna with terminal parts missing. This is a newly recorded species from the Sydney Coalfield.

Asterotheca sp. indet. (F-289, F-415).

Plate 71, Fig. 3; Plate 72, Fig. 3

SAMPLE LOCATIONS:
F-289: Harbour seam at Lingan Mine dump.
F-415: Bonar seam at Point Aconi.

Eupecopteris (Asterotheca) cyathea (Schlotheim) [Kidston, v.2, Pt. 5, p. 488, Pl. CXV] (F-14, F-409).

Plate 72, Fig. 1; Plate 73

SAMPLE LOCATION:
F-14: Unspecified coal measure of the Morien series, Cape Breton Island.
F-409: Bonar seam of Point Aconi.

BELL'S RANGE: Pt-zone (Lloyd Cove (=Lower Bonar) seam, Table 6).

REMARKS:
F-14 is a 7 cm frond with ultimate parts missing.
F-409 is a 9 cm pinna in good condition.

Pecopteris (Asterotheca) acadica Bell (F-47, F-162, F-163, F-167, F-168, F-169, F-202, Block F-248, F-284, F-320, F-322, F-397, F-399-400, F-457, F-535, F-566, 967G143.18, 967G10.22).

Plate 72, Fig. 2; Plates 74-77; Plate 78, Fig. 1

FOSSIL ASSOCIATIONS:
F-168: with *Asterophyllites* sp. indet.
F-397: with *Ptychocarpus unitus* (Brongniart).
Block F-248: with *Annularia equisetiformis* (Sternberg).
F-162: with *N. scheuchzeri* Hoffmann, Emery seam.
F-164: with *N. (Mixoneura) ovata* Hoffmann, Emery seam.
F-121: with *Annularia stellata* (Schlotheim), Emery seam.
F-401: with impressions of large sphenopterid stalk, #26 Mine dump, Harbour seam.

SAMPLE LOCATIONS:
143.18: Below Phalen seam, south shore of Indian Bay, a little east of Lingan Beach near Dominion #1 Colliery.
F-162, F-163, F-167, F-168, F-169, F-457, F-535, F-566: Emery seam.
F-47: Harbour seam, #12 Mine dump.
F-397, F-399,-400: Harbour seam, #26 Mine dump.
F-320, F-322: Harbour seam, Lingan Mine dump.
Block F-248: Stubbart seam, Prince Mine.
F-284: Stubbart seam at Point Aconi.
F-202: Unknown location in the Morien series, Cape Breton Island.
10.22: McLellan Brook, near Stellarton, N.S., Pictou series.

BELL'S RANGE: Morien series.

REMARKS:
The specific identity of F-47 is questionable. F-284 is the terminal portion of a pinna, while F-162 is a frond with terminal parts. F-202 is on permanent loan to the College of Cape Breton from Dr. G. MacLeod of Sydney, N.S. F-535 is a fertile frond of 30 cm length. *Asterotheca miltoni* (Artis) is a synonym of *P. (Asterotheca) acadica* (Bell, 1962, p.32).

Pecopteris (Asterotheca) hemitelioides Brongniart (F-201, F-203, F-238, F-239, F-285, F-285-2, F-383, F-407, F-581).

Plate 78, Figs. 2, 3; Plate 79

FOSSIL ASSOCIATIONS:
F-201: with *Asterophyllites* sp. indet.
F-203: with *Alethopteris davreuxi* (Brongniart) and *Asterophyllites equisetiformis* (Sternberg).
F-285, F-285-2: with *L. obliqua* (Bunbury), *L. bunburii* Bell and *Pecopteris (Asterotheca) acadica* Bell.

SAMPLE LOCATIONS:
F-383, F-581: Emery seam, Glace Bay.
F-238, F-239, F-285, F-285-2: Stubbart seam, Prince Mine.
F-407: Lower (?) Bonar seam at Point Aconi.
F-201, F-203: Unspecified coal measure in the Morien series, Cape Breton Island.

BELL'S RANGE: L-Pt-zones.

REMARKS:
F-285-2 is a small terminal pinna of 2 cm; F-203 is a 5 cm terminal pinna. F-201 and F-203 are on permanent loan from Dr. G. MacLeod. F-407 is several longitudinal complete pinnae of about 9 cm. The remaining specimens are partial pinnae. F-238, F-239 contain some fertile specimens.

Asterotheca robbi Bell is a synonym of *P. (Asterotheca) hemitelioides;* Corsin (1951, p. 345, Pl. 183) and Bell (1962, p. 34).

Asterotheca robbi was named by Bell (1938, p. 74-75) after Charles Robb for his valuable services to geologists and coal mine operators; see Robb (1876).

Pecopteris (Senftenbergia) pennaeformis Brongniart (F-495, F-589, F-618).

Plate 80

FOSSIL ASSOCIATIONS:
F-495: with *Mariopteris latifolia* (Brongniart), F-494.
F-618: with *Alethopteris serli* (Brongniart) and *Ptychocarpus unitus* (Brongniart).

SAMPLE LOCATIONS:
F-589: Emery seam, Glace Bay.
F-495: Harbour seam, Lingan Mine dump.
F-618: Stubbart seam, Prince Mine.

BELL'S RANGE: L-zone (Tracy and McLean seams, Table 6).

REMARKS:
F-589 is an upper part of a frond. *Senftenbergia pennaeformis* Brongniart is a synonym of *P. (Senftenbergia) pennaeformis;* see Bell (1962, p. 35).

Pecopteris sp. indet.

Plate 81

SAMPLE LOCATION:
Cape Breton Island, Morien series.

REMARKS:
On display in the Nova Scotia Museum, Halifax, N.S. This is the Carboniferous equivalent of a present-day fiddlehead, i.e., a young frond in the process of unfolding. See Gothan and Weyland (1973, p. 217, Plate 16, Figs. 5-6) for a specimen, *P. plumosa* Artis, resembling ours.

Ptychocarpus unitus (Brongniart) (F-237, F-393-395, F-398, F-398-1, F-402-1, F-487, F-540, F-584, F-607-612, F-616, F-617, 967G10.18, 967G123.28).

Plates 82-84; Plate 85, Fig. 1

FOSSIL ASSOCIATIONS:
F-237: with *Sphenophyllum cuneifolium* (Sternberg), and *P. (Asterotheca) hemitelioides* Brongniart (see also F-398).
F-487: with *Annularia sphenophylloides* (Zenker).
F-616, F-617: with *Linopteris obliqua* (Bunbury).

SAMPLE LOCATIONS:
F-540, F-584: Emery seam.
F-398-1 (GSC locality number: 508, 3820, Bell, 1938, p. 109): Backpit seam.
F-393-395, F-398: Harbour seam of #26 Mine dump.
F-487: Harbour seam, Lingan Mine dump.
F-237, F-607-612, F-616, F-617: Stubbart seam, Prince Mine.
F-402-1: Unspecified coal measure at Point Aconi.
10.18, 123.28: Morien series, Cape Breton Island.

BELL'S RANGE: Pt-zone.

REMARKS:
F-402-1 was donated to the collection by David Tracz, Sydney, N.S., and F-398-1, with Dr. Bell as collector and determiner, was donated to the collection by an anonymous private source.

F-584, with synangia (?) arranged in a single row on each side of the midrib, resembles this species; see Arnold (1947, p. 191, Fig. 87). Lately we have collected large numbers of this species as detached pinnae on one slab of rock from the Prince Mine at Point Aconi, F-607-612. Two types of venation schema are obvious: the usual number of pairs of laterals about the midrib, about four not counting the forked terminals, and in larger pinnules six to eight pairs, not counting the forked terminal vascular strands. In addition, one pinna, Pl. 82, Fig. 3, is best described as form "angustifolia," i.e., with complete fusion of pinnules such that vascular strands about the stout midrib vary from four pairs in the lower part of the pinna to a unilateral strand at the apex. This arrangement and variation over the observed length of the pinna is different from the confluent pinnules which carry the characteristic venation pattern to the tip; see Bell's description (1938, p. 77). See illustrations of venations in Fig. 7. An additional feature worth noting in specimens F-607 to 612 is that the cuticle is probably preserved.

We record for the first time from the Sydney Coalfield the physical associations of this species with *Linopteris obliqua,* F-616, and F-617.

▶ F-402-1 and F-607 to 612 may be cited as forma *emarginatus* Goeppert thus reporting this form for the first time from Sydney Coalfield; compare to Kidston's figures (1925, v. 2, Pt. 6, Pl. 81, Figs. 1; 6, 6a; 7, 7a; 9, 9a).

FILICINEAE

Pecopteris-Ptychocarpus Group of marattiaceous ferns

Fig. 8 shows several venation patterns in pinnules all from the Emery seam, bootleg pit, Glace Bay, N.S.

According to Dr. W. N. Stewart (*in litt.,* Nov. 1976), "the structures associated with the vascular strand terminals compare to linear synangia or sporangial clusters of marattiaceous fern." The mode of preservation of the structures is by ferric iron replacement without carbon remains. Magnification of 31 times does not reveal any individual sporangia; perhaps the structures represent covering membranes of sporangia (indusia).

▶ *Eoangiopteris andrewsii* Mamay is a possible identification (Andrews, 1961, p. 98-99). Compare with Fig. 1. Not previously reported from Sydney Coalfield.

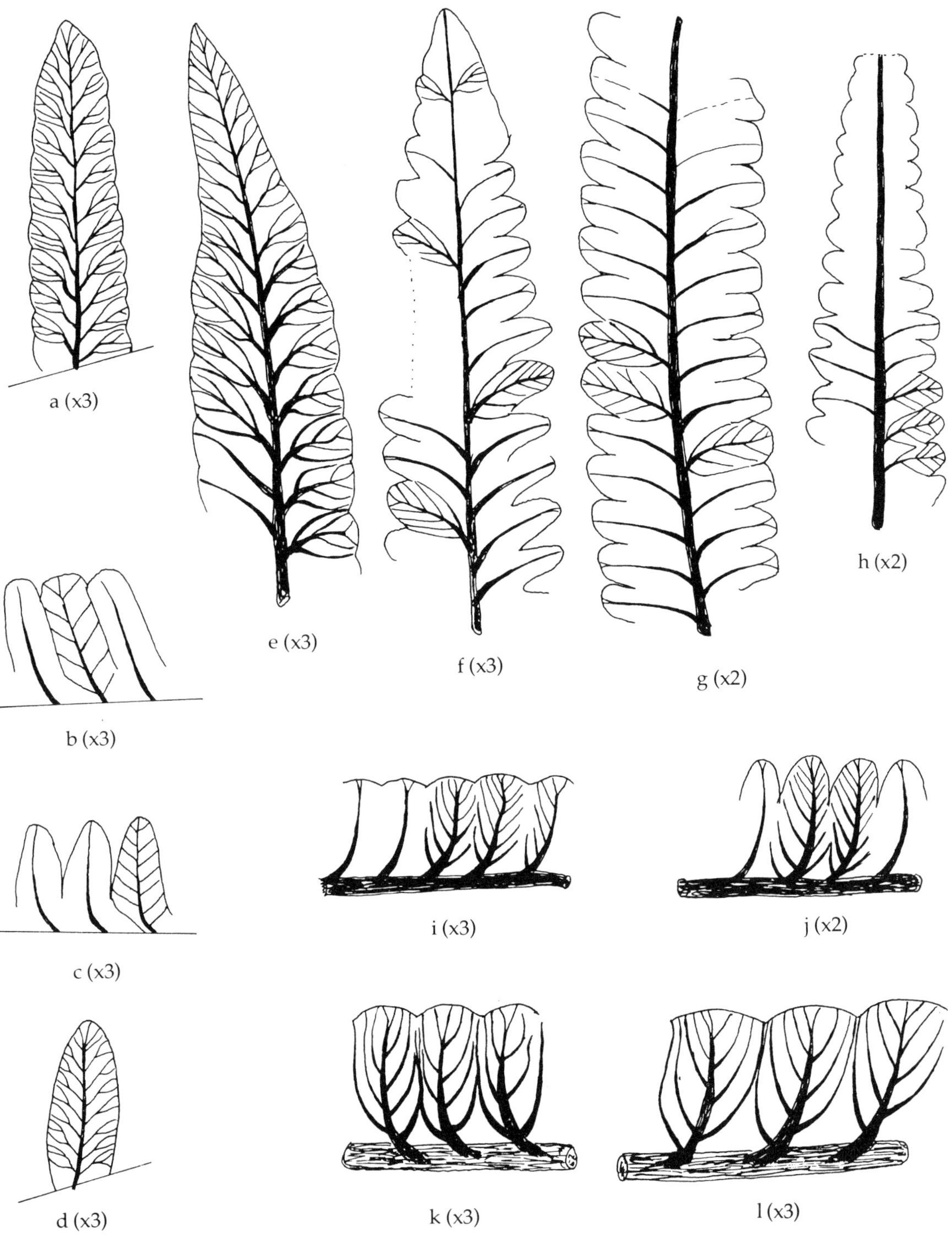

FIGURE 7
Partially drawn venation schemes characterizing the species *Ptychocarpus unitus* and pinnule morphology of its forma *emarginatus* Goeppert; a, part of pinna which is situated below d on a frond, Stephanian at Ottweiler; b, part of a pinna, Stephanian at Ottweiler; c, part of a pinna whose exact location on a frond is unknown; d, sub-terminal pinna situated above a on a frond. Corsin (1951, p. 350) called specimens a through d *Pecopteris unita* Brongniart.
Note that the positions on a frond of specimens e through l are not known; e, upper part of a pointed pinna, forma *emarginatus* (F-612), Stubbart seam; f, upper part of a pinna (F-487), Harbour seam; g, fragment of a pinna from the Mazon Creek flora, Pennsylvanian age, Illinois, USA; h, fragment of a pinna, Morien series, exact stratigraphic position is unknown; i and j are fragments of pinnae (F-608 and F-617, respectively), Stubbart seam, i is of the emarginated form; k, fragment of a pinna from an unknown stratigraphic position at Point Aconi, forma *emarginatus* (F-402-1); l, fragment of a pinna, forma *emarginatus* (F-398-1), Backpit seam.
Source: a, b, c, and d Corsin (1951, Atlas Pl. 189, Fig. 1a; Pl. 191, Fig. 2; Pl. 189, Fig. 2a; Pl. 189, Fig. 1b, respectively).
e-l are in the present collection.

FIGURE 8
Venation patterns: a is the terminal part of a pinna close to the tip of the frond; b is a portion of a pinna centrally located on a frond; and c is the terminal part of a centrally located pinna. See page 62. *Eoangiopteris andrewsii* Mamay is a possible identification.

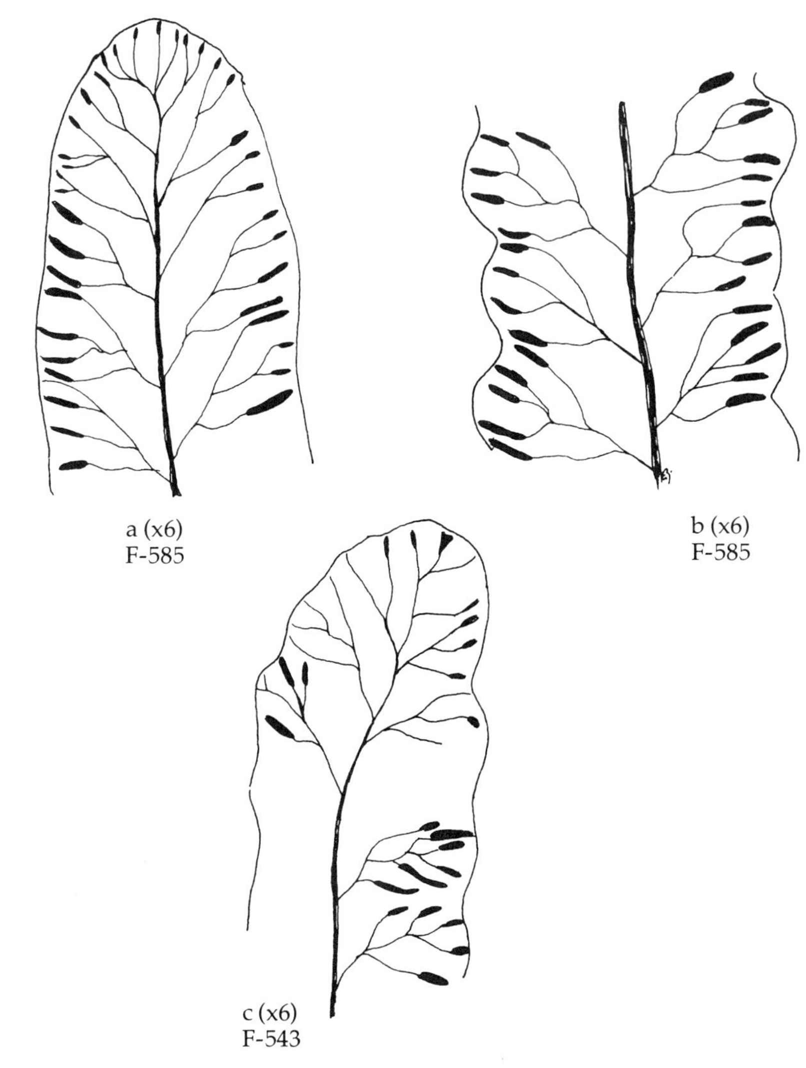

Senftenbergia sp. indet. (F-173).

Plate 85, Fig. 2

SAMPLE LOCATION:
Emery seam.

REMARKS:
Its simple or once-forked venation pattern and the shape of the pinnules suggest that this specimen be placed in this genus.

SPHENOPTERIDES

Oligocarpia missouriensis D. White (919G1.5, F-174, F-562).

Plate 86; Plate 87, Fig. 1

SAMPLE LOCATIONS:
1.5: Morien series, Cape Breton
F-174, F-562: Emery seam.

BELL'S RANGE: L-Pt-zones.

REMARKS:
The identities of specimens 1.5 and F-174 are disputable. In 1954 the genus *Oligocarpia* Goeppert was revised and a new description of this species published (Abbott, 1954, p. 51-53); see also *O. brongniarti* in Appendix II, p. 110. Formerly designated as *Sphenopteris moriensis* Bell (Bell, 1962, p. 16-17).

For additional references on this genus from the Stephanian and Permian consult Grauvogel *et al.* (1975), who discussed two fertile fronds, and Fefilova (1968) who wrote about ferns from the Ural Mountains.

Sphenopteris (Oligocarpia?) crenatodentata Bell (F-191, F-440).

Plate 87, Fig. 2; Plate 88, Fig. 1

SAMPLE LOCATIONS:
F-191: Phalen seam.
F-440: Harbour seam at Lingan Mine dump.

BELL'S RANGE: Pt-zone.

REMARKS:
F-191 is a fragment of a frond of 4 cm, while F-440 is a 5 cm frond with ultimate parts missing. There are some doubts about the identity of F-191. This species is rare in Sydney Coalfield.

Sphenopteris sp. indet. (F-290).

Plate 89, Fig. 1

SAMPLE LOCATION:
Harbour seam at Lingan Mine dump.

REMARKS:
Fertile specimens may be placed in the genus *Oligocarpia* (Abbott, 1954).

MARATTIACEAE

▶ *(?) Psaronius* sp. indet. (F-179)

SAMPLE LOCATION:
Phalen seam.

REMARKS:
There exists considerable doubt about the botanical identity of this specimen; the genus *Psaronius* is suggested (by Arnold). Perhaps the designation "gen. indet." is appropriate.

Psaronius is a large tree-fern believed to have borne pecopterid foliage. Classification is difficult (Arnold, 1947, p. 194), but Andrews (1961, p. 93-99) puts this genus, with class and family, under the ferns *(Pterophyta)*. This genus has not been previously reported from Sydney Coalfield.

GYMNOSPERMAE

PECOPTERIDES & SPHENOPTERIDES

▶ *Alethopteris* sp. indet. (F-326).

Plate 88, Figs. 2, 3

FOSSIL ASSOCIATIONS:
With *N. scheuchzeri* Hoffmann, F-324, and *Sphenopteris (Diplotmema) whitii* Bell, F-325.

SAMPLE LOCATION:
Harbour seam, Lingan Mine.

REMARKS:
Probably a crushed seed case borne by an alethopterid. The specimen is 5 cm along the "hinge line". Probably not reported previously from the Sydney Coalfield.

Dicksonites pluckeneti (Schlotheim) (967G123.17).

Plate 89, Fig. 2

SAMPLE LOCATION:
Morien series, Cape Breton Island.

REMARKS:
A 7 cm apical part of a frond.

▶ *Sphenopteris* cf. *hoeninghausi* Brongniart (F-190).

Plate 90

SAMPLE LOCATION:
Phalen seam.

REMARKS:
Arnold definitely associated this species with the frond of the pteridosperm *Lyginopteris oldhamia,* while Andrews (1961, p. 129-137) is cautious, pointing out that foliar characteristics of *L. oldhamia* appear identical to those of *S. hoeninghausi,* at least at the British Ganister coal bed. Moreover, Andrews (p. 130) placed this species in the family Lyginopteridaceae; see also Gothan and Weyland (1973, p. 292). The present specimen is 5 cm of a frond with ultimate parts missing. The botanical name *Lyginopteris (Sphenopteris) hoeninghausi* is used by Gothan and Weyland (p. 295). A species newly recorded from Sydney Coalfield.

EQUISETINAE

CALAMARIACEAE

Annularia mucronata Schenk (F-510, F-511, F-513, F-515).

Plates 96, 97

FOSSIL ASSOCIATION:
F-513: with a slender 10 cm long specimen of *Cyclopteris fimbriata* Lesquereux; the tip is present.

SAMPLE LOCATION:
All specimens were collected from the Emery seam.

BELL'S RANGE: L-Pt-zones.

REMARKS:
Bell (1938, p. 85) called this species *A. stellata* (Schlotheim) Wood forma *mucronata* Schenk.

This species is differentiated from *Annularia stellata* (Schlotheim) by the size and by the location of the widest part of the leaf; both species have mucronate structure. The leaves of the species are widest near the apex, giving rise to a spatulate to rounded apex. Those of *A. stellata* are widest near the middle of the leaves. *A. stellata* ranges from 14 mm to 75 mm in length, while *A. mucronata* is between 4 and 25 mm (Abbott, 1958, p. 307, Chart 2). Abbott believed that Bell (1938, p. 85) had both *A. stellata* and *A. mucronata* under his forma (Abbott, p. 324).

Bell's form is not recognized as a species by Abbott and she applied the above criteria to separate them as either *A. stellata* or *A. mucronata*. We follow her example.

Annularia radiata (Brongniart) (F-24, F-25, F-43, 967G10.37).

Plate 91

SAMPLE LOCATIONS:
F-24, F-25, F-43: Harbour seam, #12 Mine dump.
10.37: Unknown location in the Morien series, Cape Breton Island.

BELL'S RANGE: Morien series.

REMARKS:
Refer to Abbott (1958, Chart 2; p. 304-326) for the revaluation of North American species of the genus *Annularia* Sternberg, 1822. The means of identifying the species include data on mucro, sheath and epidermal pattern, stomata and hairs which were obtained by employing transfer methods.

Annularia sphenophylloides (Zenker) (F-442, F-488, F-601).

Plates 92, 93

FOSSIL ASSOCIATIONS:
F-422: with *Sphenopteris neuropteroides* (Boulay), F-428, F-429.

SAMPLE LOCATIONS:
F-422, F-488: Harbour seam at Lingan Mine.
F-601: Point Aconi, unknown coal measure.

BELL'S RANGE: Morien series.

REMARKS:
Block F-487, *Ptychocarpus unitus* (Brongniart), displays a multitude of this species. F-601, donated to the Collection by David Tracz, Sydney, N.S., consists of two whorls of leaves.

Annularia sp. indet. (F-26).

FOSSIL ASSOCIATION:
F-26: with *Sphenophyllum emarginatum* (Brongniart).

SAMPLE LOCATION:
F-26: Harbour seam, #12 Mine dump.

Annularia stellata (Schlotheim) (F-40, F-137, F-200, 967G10.49).

Plates 94, 95

SAMPLE LOCATIONS:
F-40: Harbour seam, #12 Mine dump.

F-137: Emery seam.
F-200: Unknown; assumed from the Morien series, Cape Breton Island.
10.49: Morien series, Cape Breton Island.

BELL'S RANGE: L-Pt-zones.

REMARKS:
Specific determination of *A. mucronata* is uncertain; see *A. mucronata* for comments on specific differentiation of the species. These specimens are probably the long-leaved variety, i.e., forma *longifolia*.

F-200 is on permanent loan to the College of Cape Breton from Dr. G. MacLeod.

Asterophyllites equisetiformis (Sternberg) (F-55, F-77, F-118, F-119, blocks F-248, F-483).

Plates 98, 99

FOSSIL ASSOCIATIONS:
F-55: with a cordaitean leaf, F-56.
Block F-483: with *N. (Mixoneura) ovata* Hoffmann, *N. tenuifolia* (Schlotheim), some cyclopterid pinnules, and *Lepidostrobophyllum* cf. *jenneyi* D. White, F-483-1.
Block F-248: with *Pecopteris (Asterotheca) acadica* Bell.

SAMPLE LOCATIONS:
F-55, F-77: Mc Aulay seam.
F-118, F-119: Emery seam.
Block F-248: Stubbart seam, Prince Mine, Point Aconi.
Block F-483: Harbour seam, Lingan Mine dump.

BELL'S RANGE: Morien series.

REMARKS:
F-119 shows seven foliar branches, two of which can be seen attached (to the calamarian stem). The specific identities of F-55 and F-77 are not certain.

It is difficult at times to distinguish this genus from *Annularia*. However, Abbott (1958, Chart 1) gave the following data for this species:

1) number of leaves per verticil is 12-20,
2) leaves linear to lanceolate, length up to 2 cm, widest in the middle,
3) width-length ratio is 1:12 to 1:20,
4) leaves are inclined 30° to 90° to the axis, and the leaf margins are convex, thus differentiating it from other species of this genus.

Note on nomenclature: the trivial name *equisetiformis* was proposed by Sternberg (not Schlotheim) (Stafleu *et al.*, 1972, *Nomina Generica Conservanda Et Rejicienda*, p. 376).

▶ *Calamites carinatus* Sternberg (F-182, F-184, F-245).

Plate 100, Fig. 1

SAMPLE LOCATIONS:
F-182, F-184: Phalen seam.
F-245: Stubbart seam, Point Aconi, Prince Mine.

REMARKS:
Weiss placed this species in the subgeneric taxon *Eucalamites* (Gothan and Weyland, 1973, p. 196).

F-245 is a 37 cm internodal type extremely abundant at the Prince Mine. The fossil is called "skidoo tracks" by miners and presents a safety hazard in the roof shale. Much longer internodal specimens were observed in that mine.

Arnold (1949, p. 181) reports a specimen of 24 cm internodal distance and 7 cm width from the Michigan coal basin. Not previously recorded from Sydney Coalfield.

▶ *Calamites cisti* Brongniart (967G10.9).

Plate 100, Fig. 2

SAMPLE LOCATION:
Morien series, Cape Breton Island.

REMARKS:
Not previously reported from the Sydney Coalfield.

▶ *Calamites cistiiformis* Stur (F-180, 967G10.40).

Plate 101, Fig. 1

SAMPLE LOCATIONS:
F-180: Phalen seam.
10.40: Unspecified coal measure, Morien series, Cape Breton Island.

REMARKS:
F-180 is a 20 cm specimen chiselled out of sandstone and reconstructed. The fossil was preserved in a position perpendicular to bedding planes. This species, generally regarded as an Upper Mississippian megaplant, has not been recorded previously from the Sydney Coalfield.

► *Calamites multiramis* Weiss (F-458).

Plate 101, Figs. 3, 4

SAMPLE LOCATION:
Emery seam.

REMARKS:
A good specimen of 35 cm. Not previously reported from Sydney Coalfield.

Calamites ramosus Artis (901G3.1, 967G143.39).

Plate 102

SAMPLE LOCATION:
3.1: one and three-quarter miles from the mouth of Halfway River at Master's Mills, Hants Co., N.S.
143.39: McLeod Brook, one mile southwest of Westville, Pictou series.

Calamites sp. (fructification) (F-252).

Plate 101, Fig. 2

FOSSIL ASSOCIATION:
With *N. (Mixoneura)* cf. *flexuosa* Sternberg.

SAMPLE LOCATION:
Stubbart seam, Point Aconi, Prince Mine.

REMARKS:
The fructification is not attached to a stem.

Calamites, sp. group? *Calamites varians* Sternberg (976GF29.1).

Plate 103, Fig. 1

SAMPLE LOCATION:
Unknown.

Calamites sp. indet. (F-199, 967G143.31).

SAMPLE LOCATIONS:
F-199: Unknown coal measure in the Morien series, Cape Breton Island.
143.31: Albion Mine quarry, Pictou Co., N.S., Pictou series.

REMARKS:
F-199 is on permanent loan to the College of Cape Breton from Dr. G. MacLeod.

Calamites suckowi Brongniart (967G13.1, 967G18.1, 967G31.2, 967G30.1, 967G143.13, 967G143.25, F-181, F-198, F-199-1, F-497, F-636, F-648).

Plate 103, Fig. 2; Plates 104-106

SAMPLE LOCATIONS:
F-181: Phalen seam.
F-636: Harbour seam, Lingan Mine dump.
143.25: Harbour seam, Sydney Mines.
F-648: Stubbart seam, Prince Mine, Point Aconi.
F-198, F-199-1: Unknown location in the Morien series, Cape Breton Island.
F-497, 31.2: Joggins, N.S.
13.1: North River Coal Mine, No. 1 seam, near McCallum settlement, Colchester Co., N.S.
18.1: Watering Brook West, West Bay, Cumberland Co., Riversdale series.
30.1: Wallace stone quarries, Cumberland Co., N.S.
143.13: Harrington River, Colchester Co., N.S.

BELL'S RANGE: Morien series.

REMARKS:
See INTRODUCTION for comments in a footnote on genus *Calamites* Suckow as "Gesamtgattung".

F-198 is a pith cast which has suffered slight vertical compression; it is 19 cm long and 6.6 cm wide.

F-198 and F-199-1 are on permanent loan to the College of Cape Breton from Dr. G. MacLeod, College of Cape Breton, and F-497 from the Beaton Institute, College of Cape Breton.

Calamites waldenburgensis Kidston (967G11.2, 967G10.41).

Plate 107, Fig. 2; Plate 108, Fig. 3

SAMPLE LOCATIONS:
11.2: Shore at McCarren's Brook, South Joggins, N.S.
10.41: Morien series, Cape Breton Island, N.S.

Macrostachya infundibuliformis (Brongniart?) (F-259, F-647).

Plate 107, Fig. 3; Plate 108, Fig. 1

FOSSIL ASSOCIATION:
With many well-preserved *Linopteris obliqua* (Bunbury) specimens.

SAMPLE LOCATION:
Stubbart seam, Prince Mine, Point Aconi.

BELL'S RANGE: Pt-zone.

Palaeostachya sp. indet. (F-269, 271, F-604).

Plate 108, Fig. 2

FOSSIL ASSOCIATIONS:
F-269, 271: with *Linopteris obliqua* (Bunbury), *Alethopteris davreuxi* (Brongniart) and *Neuropteris (Mixoneura) flexuosa* Sternberg.

SAMPLE LOCATION:
Stubbart seam, Prince Mine, Point Aconi.

BELL'S RANGE: L-zone.

REMARKS:
Bell reported his species *P. elongata* from the McLean seam (directly above the Tracy seam, Table 7).

SPHENOPHYLLINAE

SPHENOPHYLLACEAE

Sphenophyllum cuneifolium (Sternberg) (F-23, F-45, F-53, F-196).

Plate 107, Fig. 1; Plate 109

SAMPLE LOCATIONS:
F-53: Mc Aulay seam.
F-196: Phalen seam, Glace Bay.
F-23, F-45: Harbour seam, #12 Mine dump.

BELL'S RANGE: Lch-L-zones.

REMARKS:
F-53 is a 6 cm specimen with 10 whorls of well preserved leaflets; the mode of preservation gives this specimen a three-dimensional aspect.

See *S. emarginatum* (Brongniart) for a note on classification.

Sphenophyllum emarginatum (Brongniart) (F-18, F-46, F-227, F-628, F-642-2, 967G10.43).

Plate 110

FOSSIL ASSOCIATION:
F-642-2 with *Linopteris obliqua* (Bunbury) and different anatomical parts (vegetative?) of the species.

SAMPLE LOCATIONS:
F-628: Phalen seam.
F-642-2: Harbour seam, Lingan Mine.dump.
F-18, F-46: Harbour seam, #12 Mine dump.
F-227: Stubbart seam, Prince Mine, Point Aconi.
10.43: Unknown location in the Morien series, Cape Breton Island.

BELL'S RANGE: L-Pt-zones.

REMARKS:
The present classification differs from Bell's (1938, p. 89) in that *Sphenophyllum* sp. with much more broadly wedge-shaped leaflets are identified as *S. emarginatum* (Dr. Arnold, *in litt.*, Aug. 24, 1977). Gothan and Weyland follow similar criteria in the classification of this species (1964, p. 160).

Excellent reference material consisting of line drawings and photographs is presented by Batenburg (1977). In addition, Batenburg in that publication contributed in a major way to the understanding of the vegetative parts of this species. Not recognized before in paleobotany, and described for the first time, were the following summarized features (Fig. 5, p. 91, 94);

1) occurrence of hairs on an axis and on lateral leaf margins;
2) young branches arising from the main axis;
3) the apex of the main axis (it does not fade into a twig-like axis with cuneiform leaves, but terminates as thinning brushlike bunches (Pl. 5, Fig. 3), and
4) uncate aristae of the leaves at the main axes and main branches may be interpreted as "barbed hooks" so that this species was a "clinging, ascending to climbing plant" supporting each other as a dense low mass or supporting itself on other species, for example on *Linopteris obliqua* plants, as block 977GF-642-2 seems to indicate.

The points necessitate an emended diagnosis which, among other things, emphasizes the heterophyllous nature of the leaves in this species.

Batenburg's material was collected from one outcrop exposed by constructing highway A1 in the Saarland, Federal Republic of Germany.

Note on nomenclature: the type specimen for the genus *Sphenophyllum* Brongniart 1828 is *S. emarginatum* (Brongniart) Brongniart (Stafleu *et al.*, 1972, p. 376); *Sphenophyllites emarginatus* Brongniart is a *nominum rejiciendum*. Emended diagnosis of *S. emarginatum* (Batenburg, 1977, p. 94):

"Heterophyllous. Leaves of the twigs (those of the basal nodes may be different) 6(-9) pro verticil, cuneiform, 4 mm to almost 2 cm long, ca. 2-10 mm wide at the distal margin; length — width ratio between 1 and 5. Lateral margins slightly convex to slightly concave; distal margin straight to slightly convex, with ca. 4-18 semicircular teeth, and acute sinuses. Teeth ca. 0.5 mm long, ca. 0.5 mm wide at the base; length — width ratio ca. 1. A single vein entering the leaf, dichotomizing close by the base, and subsequently 1-4 times, with generally increasing intervals towards the distal margin. Distal margin often (especially in the broader leaves) with a shallow median cleft; secondary clefts may occur in the resulting lobes. Leaves smaller at the apices of the twigs. Leaves at the basal nodes of the twigs transitional between (and often smaller than) those at the corresponding node of the relative main branch on the one side, and the leaves at the rest of the twig on the other.

"Leaves at the larger branches ca. 0.5-1.5 cm long. They are either undivided cuneiform and resembling those at the twigs (6 leaves pro verticil); or broader and deeply dissected, with ca. 8 linear, nearly equal lobes (6 leaves pro verticil); or narrow and more or less deeply 2-, 3-, or 4-lobed (6-9 leaves pro verticil); or linear to linear — lanceolate (12 leaves pro verticil); or transitional between these types. Leaf types occurring in different sequences along the branches. Linear leaves and lobes of deeply dissected leaves ca. 0.5-1 mm wide in their widest part, gradually narrowing into an uncate arista. Some leaves provided with hairs at the lateral margins.

"Leaves of the main axes ca. 0.5 (apical ones) -1.5 cm long. They are either linear (12 leaves pro verticil); or narrow and deeply bipartite (6 leaves pro verticil); or broader and deeply dissected, with several linear, nearly equal lobes (6 leaves pro verticil). Linear leaves and lobes of the other leaves ca. 0.5-1.5 mm wide in their widest part, gradually narrowing into a strongly uncate arista.

"Internodes of the main axes up to 4 cm long (towards the apex strongly shortening to less than 2 mm), ca. 3.0-6.5 mm wide, ribbed. Internodes of the large branches up to over 3 cm long, ca. 1.2 cm to over 3 mm wide, with either 6 or 3 ribs. Internodes of the twigs ca. 2.5-15 mm long (shortening both towards the apex and the base of the twigs), ca. 0.5-1.3 mm wide, with 3 ribs. Nodes either not, or slightly, or strongly swollen. Some of the axes provided with perpendicular hairs; those of the main axes, if any, concentrated at the upper half of the internodes. Hairs up to 1 mm long, up to 0.1 mm in diameter at the base. Generally only a single branch at the node; sometimes 2. Intervals between subsequent branches varying. Branches may be spirally arranged (dextrorsally)."

Compare this to Storch (1966). Fructifications were not found on the specimens, so refer to Storch for emendation.

Sphenophyllum majus (Bronn) (F-297, F-298, F-302, F-303).

Plate 111

FOSSIL ASSOCIATIONS:
F-303: with *L. dawsoni* Bell, and *S.* sp. indet. (stem), F-301.

SAMPLE LOCATION:
Harbour seam, Lingan Mine.

BELL'S RANGE: Pt-zone.

REMARKS:
See note on classification under *S. emarginatum* (Brongniart).

Sphenophyllum oblongifolium (Germar and Kaulfuss) (967G10.42, 967G10.54, F-370, F-481).

Plate 112

SAMPLE LOCATIONS:
10.42: Morien series, Cape Breton Island.
10.54: Morien series, Cape Breton Island.
F-370: Harbour seam, Lingan Mine.
F-481: Phalen seam.

BELL'S RANGE: Pt-zone.

REMARKS:
See note on classification under *S. emarginatum* (Brongniart).

Sphenoplyllum sp. indet. (F-175, F-301).

Plate 113, Fig. 1

FOSSIL ASSOCIATIONS:
F-301: with *L. dawsoni* Bell and *S. majus* (Bronn).

SAMPLE LOCATIONS:
F-175: Emery seam, F-301: Harbour seam, Lingan Mine.

Sphenophyllum sp. indet. (fertile) (F-88, F-91).

Plate 113, Figs. 2, 3

SAMPLE LOCATION:
Shoemaker seam.

REMARKS:
F-88 resembles a cone.

Sphenophyllum cf. *trichomatosum* Stur (F-63).

Plate 114, Fig. 1

SAMPLE LOCATION:
Mc Aulay seam.

BELL'S RANGE: L-Pt-zones.

LYCOPODINAE

LEPIDODENDRACEAE*

*Abbott (1963) placed this Family under the Class Lycopsida which is subordinated to the Subdivision Paleocormophyta.

▶ *Lepidodendron aculeatum* Sternberg (967G10.47, 967G10.65, 967G10.85, F-288).

Plate 114, Fig. 2; Plate 115

FOSSIL ASSOCIATIONS:
10.85; with *Neuropteris scheuchzeri* Hoffmann.
F-288 verso: with *Linopteris obliqua* (Bunbury), and *Eremopteris artemisiaefolia* (Sternberg).

SAMPLE LOCATIONS:
10.47: Coal mine north of Mabou Harbour, Cape Breton Island.
10.65: Little Glace Bay, Harbour seam.
10.85: Unspecified coal seam, Morien series, Cape Breton Island.
F-288: Harbour seam, Lingan Mine dump.

REMARKS:
Bell (1944) is uncertain about his collected specimens; see also Bell (1966, p. 26, Plate XII, Fig. 2).

This family contains two genera, *Lepidodendron* Sternberg and *Lepidophloios* Sternberg (Table 8) (including their organ-genera, branch-genera, and one common genus for the root system). Not previously reported from Sydney Coalfield.

Lepidodendron bretonense Bell (967G10.88, F-29, F-294, F-505, F-651, F-656).

Plates 116, 118

SAMPLE LOCATIONS:
F-505: Phalen seam.
F-29: Harbour seam, # 12 Mine dump.
F-294: Harbour seam, Lingan Mine.
F-651, F-656: Stubbart seam, Prince Mine.
10.88: Cape Breton Island, Morien series.

BELL'S RANGE: L-Pt-zones.

REMARKS:
F-29 is a finely-preserved small specimen. F-294, the identity of which is disputable, is partially replaced by pyrite. *Lepidodendron dichotomum* var. *bretonensis* Bell is a synonym of *L. bretonense;* see Bell (1962, p. 53).

Lepidodendron dawsoni Bell (F-28, F-97b, F-299, F-496-1-3).

Plate 119

FOSSIL ASSOCIATIONS:
F-97b: with *Sphenopteris striata* Gothan, F-99-100.
F-299: with *Sphenophyllum* sp. indet., F-301; and *Sphenophyllum majus* (Bronn), F-303.

SAMPLE LOCATIONS:
F-97b: Shoemaker seam.
F-28: Harbour seam, #12 Mine dump.
F-299, F-496-1-3: Harbour seam, Lingan Mine dump.

BELL'S RANGE: Pt-zone.

REMARKS:
F-97b is a possible ultimate branch of an arborescent lycopod, i.e., *L. dawsoni* or *Lycopodites meekii* Lesquereux.

▶ *Lepidodendron ophiurus* Brongniart (967G10.44).

Plate 120, Fig. 2

FOSSIL ASSOCIATION:
With *Neuropteris (Mixoneura)* cf. *ovata* Hoffmann.

SAMPLE LOCATION:
Morien series, Cape Breton Island.

REMARKS:
A 7 cm branching stem. Not previously reported from Sydney Coalfield.

Lepidodendron pictoense Dawson (F-436, F-507, F-652, F-670).

Plate 117; Plate 120, Fig. 3

SAMPLE LOCATIONS:
F-507, F-670: Phalen seam, Glace Bay.
F-436: Harbour seam, Lingan Mine dump.
F-652: Stubbart seam, Prince Mine, Point Aconi.

REMARKS:
L. lycopodioides Sternberg is a synonym of *L. pictoense;* see Bell (1962, p. 52).

Lepidodendron sp. indet. (F-354, 967G36.11).

Plate 121

SAMPLE LOCATIONS
Harbour seam, Lingan Mine for F-354, and an unspecified coal seam in the Morien series for 36.11.

Lepidodendron sp. (leafy stem) (F-45-1).

Plate 120, Fig. 1

FOSSIL ASSOCIATION:
With *Sphenophyllum cuneifolium* (Sternberg).

SAMPLE LOCATION:
Harbour seam, #12 Mine dump.

Lepidodendron sp. (twig) (F-89, F-96, F-97, F-467).

Plate 122

SAMPLE LOCATION:
All specimens came from the Shoemaker seam.

REMARKS:
This is a good locality for the collection of this species. F-97 is partially replaced by pyrite. Some twigs are up to 20 cm long.

Lepidodendropsis corrigatum (Dawson) (967G35.1).

Plate 123, Fig. 1

SAMPLE LOCATION:
Lower Carboniferous, Horton, N.S.

Lepidophloios laricinus Sternberg (967G10.15, 967G31.4, 967G36.12, 967G123.20, 967G143.36, 967G123.1, 12).

Plate 123, Figs. 2, 3; Plate 124, Fig. 1

SAMPLE LOCATIONS:
10.15, 36.12, 123.20, 123.1, 12: Morien series, Cape Breton Island.
31.4: South Joggins, Cumberland Co., N.S.
143.36: North Slope, Springhill, Cumberland Co., N.S.

REMARKS:
123.1, 12 are forma *Halonia tortuosa* Lindley and Hutton, i.e., they have raised scars showing spiral rather than vertical sequence.

▶ *Lepidophyllum* sp. (F-411).

Plate 124, Fig. 2

FOSSIL ASSOCIATIONS:
With *N. scheuchzeri* Hoffmann, and cyclopterid pinnules.

SAMPLE LOCATION:
Upper? Bonar seam, Point Aconi.

REMARKS:
The 4 cm specimen, believed to be the apical part of the leaf, is slightly curvi-triangular, with two prominent striae in the center which converge at the apex.

This genus has probably not been reported from the Sydney Coalfield.

Lepidostrobophyllum jenneyi (D. White) (F-483-1, block F-623, F-655).

Plate 125

FOSSIL ASSOCIATIONS:
F-483-1 on block F-483: with *Asterophyllites equisetiformis* (Sternberg), and some cyclopterid pinnules.
Block F-623: with *Sigillariophyllum* sp., *Sphenophyllum* sp. and *Lepidostrobophyllum* sp.

SAMPLE LOCATIONS:
F-483-1: Harbour seam, Lingan Mine dump.
F-655: Stubbart seam, Prince Mine.
Block F-623: Phalen seam, 1-B Mine, #26 Colliery.

BELL'S RANGE: L-Pt-zones.

REMARKS:
F-655 is a conal fragment with triangular sporophylls of the *Lepidostrobophyllum jenneyi* type. This fragment is 15 cm long. Specimen F-483-1 and Block F-623 are sporophylls which conform to the description of this species.

Lepidostrobophyllum lanceolatum (Lindley and Hutton) (967G10.21, 967G20.10, F-642).

Plate 126, Figs. 2, 3

SAMPLE LOCATIONS:
F-642: Harbour seam, Lingan Mine dump.
10.21: Morien series, unknown location, Cape Breton Island.
20.10: McLellan Brook, two miles east of Stellarton, Stellarton series.

REMARKS:
Differentiation between detached sporophylls and complete cones is required; see Crookall (1966, v. 4, Pt. 4, p. 503-504). Accordingly 967G10.21 and F-642 are called *Lepidostrobus lanceolatus* Lindley and Hutton because these specimens are cone compressions, while the species' name applies to 967G20.10 because the specimen consists of detached sporophylls. The former has not been recorded from Sydney Coalfield. Abbott (1963) placed *L. lanceolatum* under *Lepidocarpopsis lanceolatus* (Lindley & Hutton) Abbott.

Lepidostrobophyllum lanceolatum (Lindley and Hutton) var. *constrictum* Bell [1938, p. 97] (F-506).

Plate 124, Figs. 3, 4

FOSSIL ASSOCIATIONS:
With species of *Lepidostrobophyllum* cf. *mintoensis* Wilson; *Sphenophyllum* sp. indet. and one (2.5 cm) pinna which shows the mixoneuroid condition.

SAMPLE LOCATION:
Phalen seam.

BELL'S RANGE: L-Pt-zones.

REMARKS:
This leaf of a cone is 3.5 cm long and about 6 mm at its widest. The constricted base, that portion which bears the sporangium, is visible together with the midvein.

Lepidostrobus cf. *mintoensis* Wilson (F-471).

Plate 127

SAMPLE LOCATION:
Phalen seam, Pt-zone.

BELL'S RANGE: L-Pt-zones.

REMARKS:
This is a specimen of a cone compressed along the axis with blades about 12 mm long, triangular or slightly convex; the blades are about 6 to 7 mm wide. The dimensions of the wedge-shaped pedicel are difficult to measure but appear to be in the neighborhood of 7 mm. Angles tend to project slightly beyond the junction of blades and pedicel. Refer to Bell's species (Bell, 1938, p. 95) which seems to compare well with our specimen.

For a discussion of chromosomes from the Pennsylvanian arborescent lycopod cone, *L. schopfii,* see Brack-Hanes and Vaughan (1978).

Lepidostrobophyllum triangulare Bell is a synonym of *Lepidostrobus mintoensis;* see Bell (1962, p. 54).

Lepidostrobus sp. indet. (901G3.21, 967G10.84, F-328, F-563).

Plate 126, Fig. 1; Plate 138, Figs. 2, 3; Plate 139, Fig. 1

SAMPLE LOCATIONS:
F-563: Emery seam, Glace Bay.
F-328: Harbour seam, Lingan Mine dump.
3.21: one and three-quarter miles from the mouth of Halfway River, at Master's Mills, Hants Co., N.S.
10.84: Little Bras d'Or, Cape Breton Island.

BELL'S RANGE: Pt-zone.

REMARKS:
F-563 is a cone, 12 cm long and 1.4 cm at its widest. The sporophylls are slender, and subtriangular with a basal width of two to four mm, and length of about 7 mm. The exposed axis in the center of the cone is between three and four mm wide, showing leaf scars arranged roughly in a vertical manner (a cone from a sigillarian, such as *Sigillariostrobus* species, is similar).

▶ *Lepidostrobus variabilis* Lindley and Hutton (872G1.4, 967G10.52).

SAMPLE LOCATIONS:
1.4: George McKay Mine, New Glasgow, N.S., Stellarton series.
10.52: Morien series, unspecified coal seam, Cape Breton Island.

REMARKS:
Apparently not previously reported from Sydney Coalfield; see Bell (1962a, p. 43).

*See Abbott 1963

SIGILLARIACEAE*

▶ *Sigillaria* cf. *brardi* Brongniart (F-350).

Plates 128, 129

SAMPLE LOCATION:
Harbour seam at Lingan Mine dump.

REMARKS:
This specimen resembles *S. brardi* and appears to belong to the group *Subsigillaria* of the *Leiodermaria* type, i.e., nonribbed and smooth-skinned. The leaf cushions in the specimen are visible only under oblique light; ridges between cushions are absent. The leaf scars are separated by about 1.1 to 2.1 cm horizontally, and 1.7 to 2.2 cm vertically. The scars are hexagonal-elongate, 1.1. to 1.2 cm across, exhibiting a tessellate structure.

Gothan and Weyland (1964, p. 134) reported that *S. brardi* is one of the youngest sigillarians, i.e., heading for extinction then, but still found in the lower Permian (Rotliegendes) in Germany. Not previously reported from Sydney Coalfield; possibly a new variety.

Sigillaria elegans (Sternberg) (967G10.81, F-292, F-349, F-433).

Plates 130, 131

SAMPLE LOCATIONS:
10.81: Morien series, Cape Breton Island?
F-292, F-349, F-433: Harbour seam, Lingan Mine dump.

BELL'S RANGE: Pt-zone (Harbour seam, Queen pit in Sydney Mines, N.S.).

REMARKS:
The identity of F-433 is uncertain; F-349 resembles this species. Weiss distinguished between *Eusigillaria* and *Subsigillaria;* the first mentioned subgenus is furthermore divided into Group *Rhytidolepis* (ribbed, channeled), and Group *Favularia* (honeycomb structure).

A conspicuous member of the favularian group is *S. elegans*. Refer to *S.* cf. *brardi* as an example of *Subsigillaria*.

Sigillaria laevigata Brongniart (F-309, F-366).

Plate 132

SAMPLE LOCATION:
Both specimens are from the Harbour seam, Lingan Mine.

BELL'S RANGE: Pt-zone (Emery seam in Glace Bay, N.S.).

REMARKS:
The identity of F-366 is not certain.

F-309 is a decorticated stem of syringodendron preservation; the natural leaf scars have disappeared by destruction of the outermost tissues, leaving large and conspicuous pairs of parichnos prints (side scars). This preservation condition is detailed in plates accompanying *Sigillaria* sp. indet., F-115, F-116, and F-352 in particular.

▶ *Sigillaria scutellata* Brongniart (967G10.80, 967G31.6, 967G20.7).

Plate 133

SAMPLE LOCATIONS:
10.80: Morien series, unspecified coal measure, Cape Breton Island.
31.6: South Joggins, Cumberland Co., N.S.
20.7: McLellan Brook, two miles east of Stellarton, N.S., Stellarton series.

REMARKS:
Not previously reported from Sydney Coalfield.

▶ *Sigillaria* sp. (F-437).

Plate 134

SAMPLE LOCATION:
Harbour seam, Lingan Mine dump.

REMARKS:
The structure is similar to that of the *Favularia* group, in that the leaf cushions without ridges are tangential, giving rise to a faviform appearance; however, this specimen lacks ribs. The cushions are about 8 mm across and have a prismatic habit similar to that of *S.* cf. *brardi*. The specimen is small and not well preserved. Possibly a new botanical species.

Sigillaria sp. indet. (967G31.3, 967G143.21, F-115, F-116, F-312, F-313, F-352, F-622-1, F-626).

Plates 135, 136

SAMPLE LOCATIONS:
F-115, F-116: Emery seam.
F-626: Phalen seam, #26 Colliery, 1-B Mine.
F-312, F-313, F-352: Harbour seam, Lingan Mine dump.
F-622-1: Stubbart seam, Prince Mine, Point Aconi.
31.3: South Joggins, N.S.
143.21: Acadia Colliery, Westville, Acadia Main Seam, Stellarton series, Pictou Co., N.S.

REMARKS:
Both F-115 and F-116 are casts, 40 cm and 35 cm in diameter, respectively, and probably two different species; they were dug out in the bootleg pit (see Fig. 11). F-115, F-116, F-312, F-352, F-622-1 and F-626 exhibit syringodendron preservation (see *S. laevigata* for an explanation of the term).

Sigillaria tessellata (Steinhauer) (861G1.1, 967G10.83, 90, 967G31.7, F-459-3).

Plate 137; Plate 138, Fig. 1

SAMPLE LOCATIONS:
1.1: Sydney Mines, Sydney Main seam?
10.83, 90: Morien series, Cape Breton Island.

31.7: Albion Mines, New Glasgow, Pictou Co., N.S.
F-459-3: Emery seam.

BELL'S RANGE: Pt-zone.

▶ *Sigillariophyllum* sp. (Block F-623, F-623-1, F-624).

Plate 144

FOSSIL ASSOCIATION:
Block F-623: with sporophyll of the *Lepidostrobophyllum jenneyi* type.

SAMPLE LOCATION:
Phalen seam, #26 Colliery, 1-B Mine (which is out of production).

REMARKS:
This species occurs over large areas in this old mine and in physical association (overprints) with *Sigillaria* species. Some of the specimens are over three feet long.

The midrib of this species is rather pronounced, resembling a straight-sided groove. This genus has not been previously reported from Sydney Coalfield.

Stigmaria ficoides (Sternberg) (967G12.1, 967G12.4, 5, F-49, F-351, F-606, F-641).

Plate 139, Fig. 2; Plate 140

SAMPLE LOCATIONS:
F-351, F-641: Harbour seam, Lingan Mine dump.
F-49, F-606: unknown location in the Morien series, Cape Breton Island.
12.1: Drummond Colliery, Westville, Stellarton series.
12.4, 5: Acadia Colliery, Westville, Acadia Main Seam, Stellarton series.

BELL'S RANGE: Morien series.

REMARKS:
12.1 is an unusual specimen because of its extreme nodular appearance. F-606, donated by B. J. D. Irwin, Port Morien, Cape Breton Island, is a specimen showing the vascular core.

Stigmaria sp. indet. (F-208, F-244, F-353, F-357, F-358).

Plates 141, 142

SAMPLE LOCATIONS:
F-353, F-357, F-358: Harbour seam, Lingan Mine dump.
F-208, F-244: Stubbart seam, Prince Mine.
For F-244 see COAL SAMPLES, Appendix I.

REMARKS:
F-353 is an uncommon type, while F-208 consists of detached (true) roots. Roots of about 10 cm or less extend into surrounding sediments from a stigmarian log, F-357. F-358 is 15 to 25 cm long and represents dichotomizing stigmarians.

GYMNOSPERMAE

CORDAITACEAE

Artisia sp. indet. (976GF80.1, 967G123.13).

Plate 143, Figs. 1, 2

SAMPLE LOCATIONS:
80.1: Unknown, assumed to be from Nova Scotia.
123.13: Port Hood, Cape Breton Island.

Artisia transversa (Artis) (967G11.1).

Plate 143, Fig. 3

SAMPLE LOCATION:
Shore at McCarren's Brook, South Joggins, Cumberland Co., N.S.

REMARKS:
This specimen is a pith cast of a cordaitean stem and is 10 cm long.

Cordaianthus sp. indet. (967G12.2).

SAMPLE LOCATION:
Acadia Colliery, Westville, Acadia Main Seam, Stellarton series, Pictou Co., N.S.

REMARKS:
This is the inflorescence of a member of the genus *Cordaites;* whether it is the male or female inflorescence cannot be determined.

Cordaites principalis (Germar) (872G2.24, 967G143.34, 967G26.1, 967G20.6).

Plate 145, Fig. 1

SAMPLE LOCATIONS:
2.24: Caledonia Slope, Cape Breton Island.
143.34: Grand Lake Road, north of Sydney, Morien series, Cape Breton Island.
26.1: Port Hood, Riversdale series, Cape Breton Island.
20.6: McLellan Brook, Stellarton series, Pictou Co., N.S.

Cordaites sp. indet. (F-65, F-355, F-56).

Plate 145, Fig. 2; Plate 146, Fig. 1

FOSSIL ASSOCIATION:
F-56: with *Asterophyllites* cf. *equisetiformis* (Sternberg), F-55.

SAMPLE LOCATIONS:
F-65, F-56: Mc Aulay seam.
F-355: Harbour seam, Lingan Mine.

REMARKS:
F-65 represents fragments of cordaitean leaves.

Cordaites sp. (leaves) (900G1.2).

SAMPLE LOCATION:
Smelt Brook, Trenton, N.S., Pictou series.

REMARKS:
The leaves are about 4 cm long.

UNKNOWN AFFINITY

▶ *Triletes* cf. *auritus* var. *grandis* Zerndt (F-120-1).

Plate 146, Figs. 2, 3

FOSSIL ASSOCIATIONS:
With *Odontopteris subcuneata* Bunbury, *Alethopteris serli* (Brongniart) and *Neuropteris scheuchzeri* Hoffmann on Block F-120.

SAMPLE LOCATION:
Emery seam, Glace Bay.

REMARKS:
There are other megaspores on this block, i.e., F-120-2, which are broken in cross-section; this makes it difficult to identify them. The specimen is 1 mm across. Generally, this genus of spores is attributed to lycopods (Andrews, 1961, p. 252).

The specific identification is disputable. This specimen, very much like the form-genera of fern-like foliage, is tentatively classified by morphology as follows:

Anteturma:	Sporites
Turma:	Zonales
Subturma:	Auritotriletes
Infraturma:	Auriculati
Genus:	*Triletes*.

For further information see Burbridge and Felix (1976, Pl. 1) and their discussion of *Spencerisporites*, a spore genus regarded as being of lycopod affinity. Not previously reported from Sydney Coalfield.

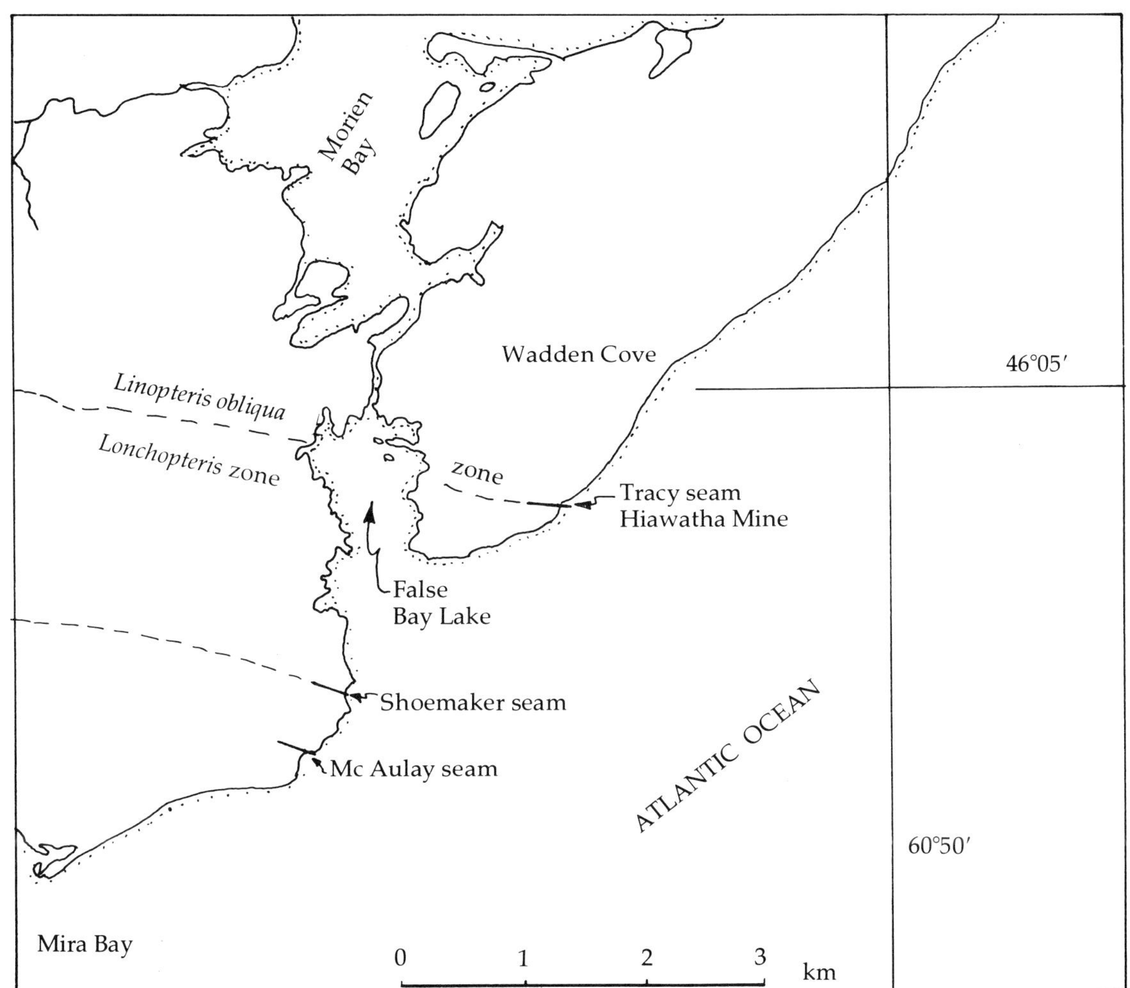

FIGURE 9
Locations of Mc Aulay, Shoemaker and Tracy seams (dashed lines) on Cape Breton Island.
(Redrawn after Geology Map 362A, Glace Bay Sheet, Cape Breton County, Nova Scotia, 1938.)

FIGURE 10
A Sigillarian species *in situ*, bootleg pit, Emery seam, Glace Bay, N.S. The distribution of aborescent lycopods of this group in the pit is shown in Fig. 11.

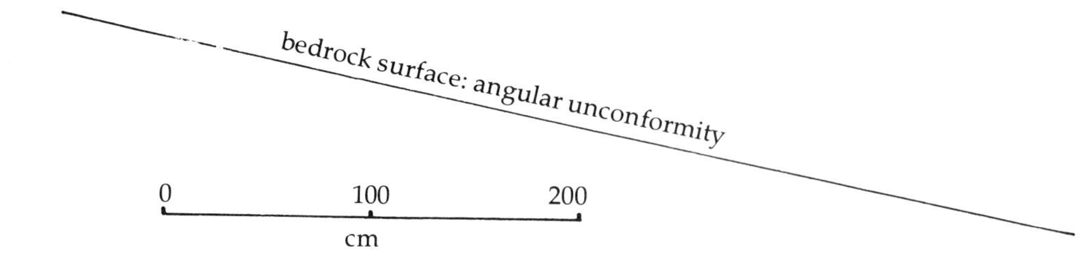

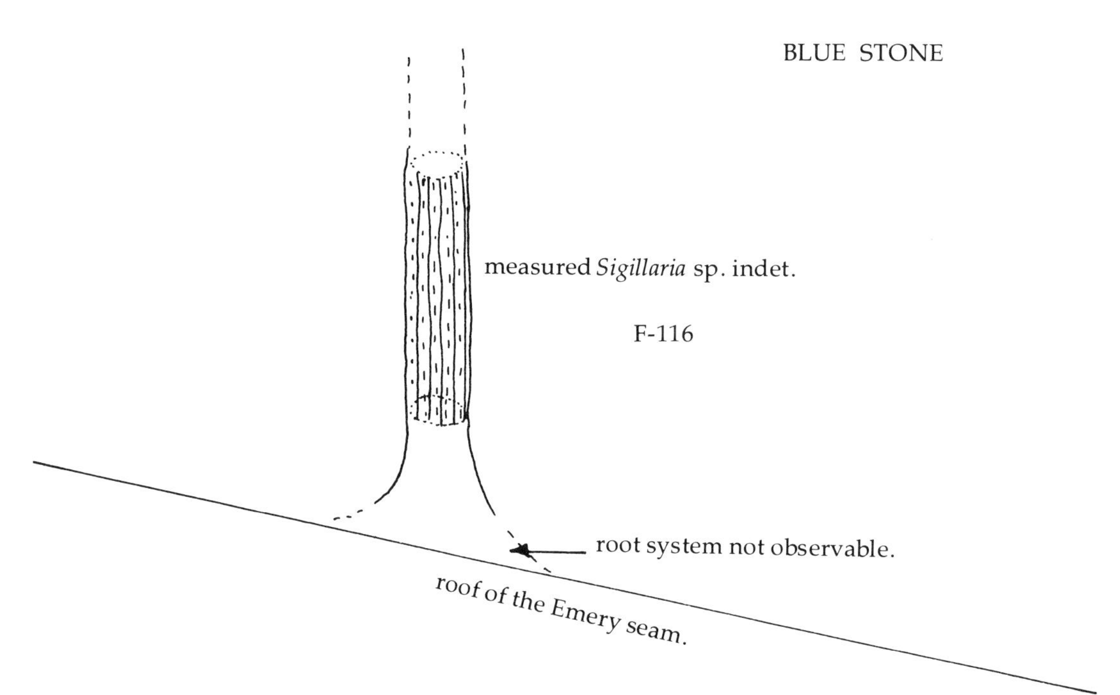

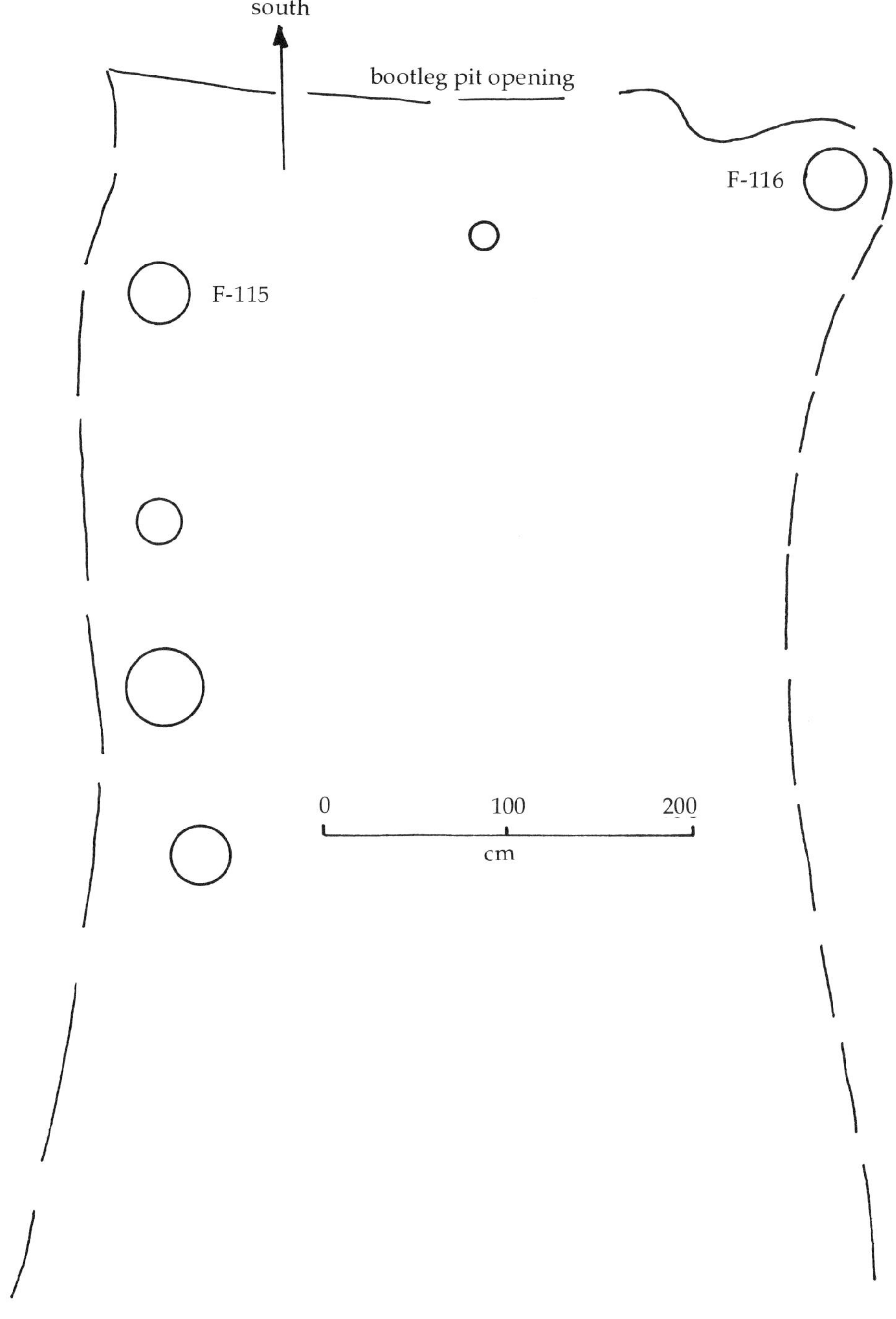

FIGURE 11
Inclined plan of the entrance to the bootleg pit in Glace Bay, N.S.

Positions of *Sigillaria* sp. indet. are shown by circles which indicate the scale of the tree casts preserved in the syringodendron condition. Entrance to the bootleg pit was on July 23, 1976. The entrance was caved-in, and the Emery seam inaccessible in June 1977. F-115 and F-116 specimens are in the present collection.

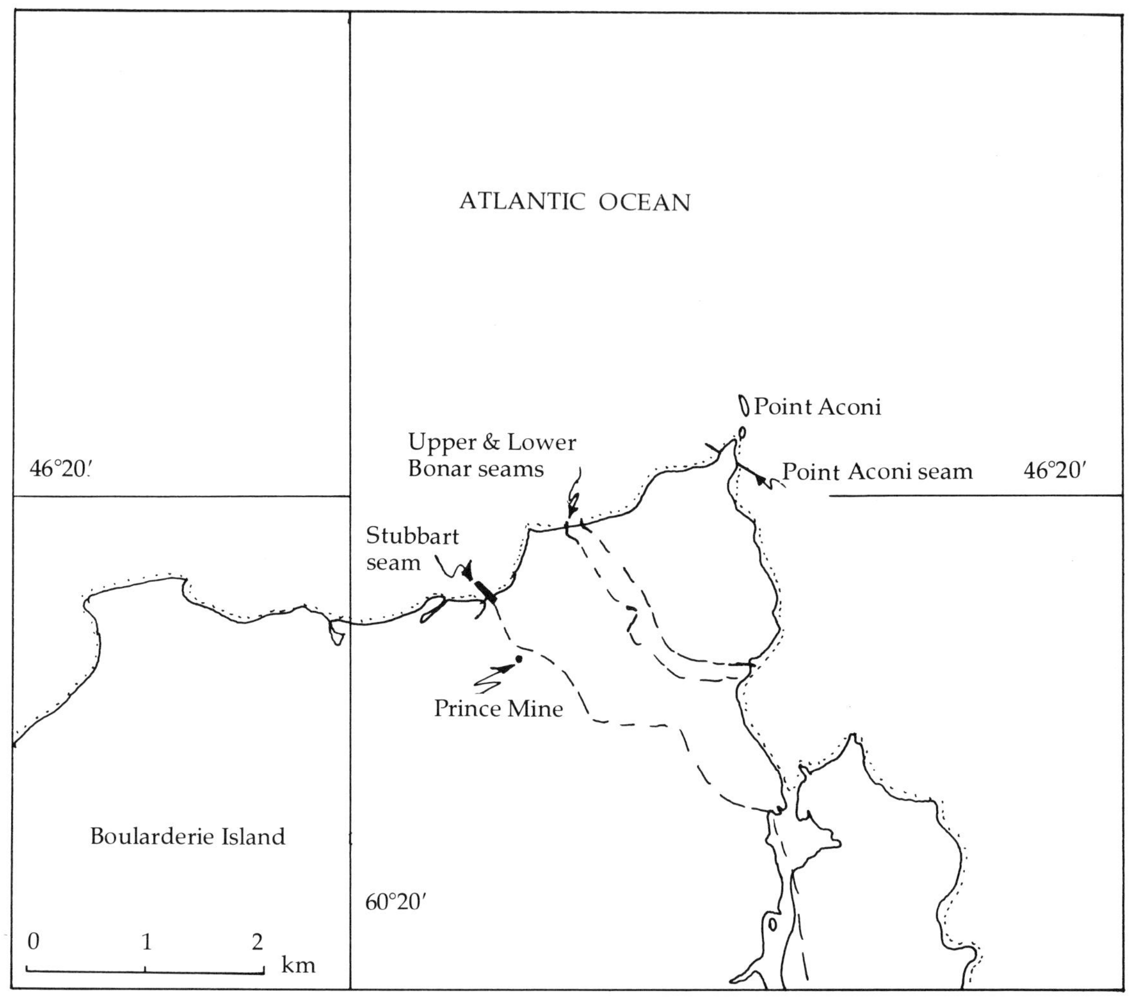

FIGURE 12
Location map of Prince Mine, Stubbart, Upper and Lower Bonar and Point Aconi seams; Point Aconi seam is the youngest outcropping coal seam in the Sydney Coalfield. Dashed lines are coal seams.
Between Stubbart and Upper Bonar seams *Odontopteris subcuneata* and *Pecopteris (Asterotheca) hemitelioides* were collected. (Redrawn from Geology Map 359A, Bras d'Or Sheet, Cape Breton and Victoria Counties, Nova Scotia, 1938.)

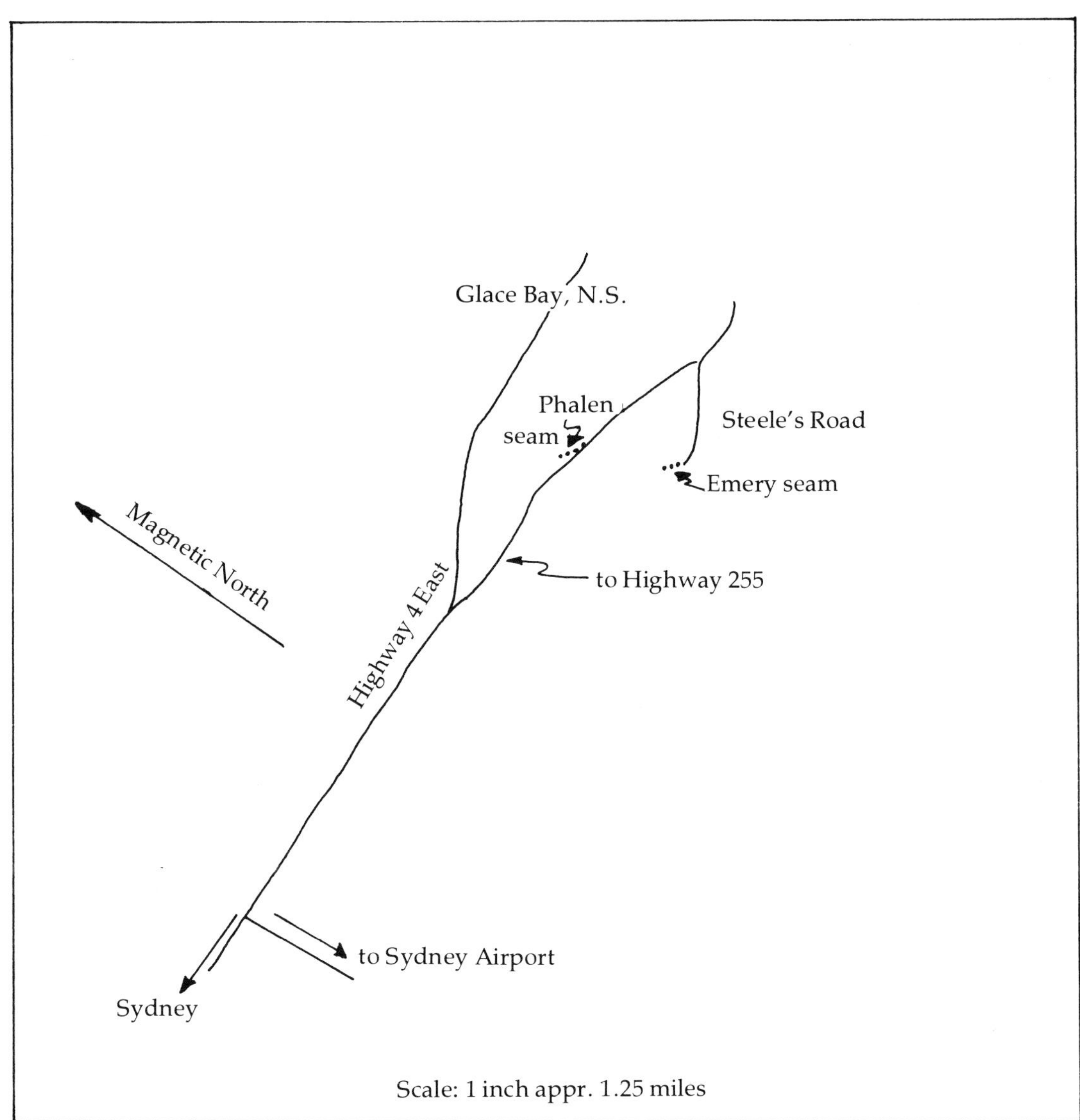

FIGURE 13
Road map to locate the outcrops of the Phalen and Emery seams in Glace Bay, N.S.

The Phalen seam outcrops (1977) directly on the left-hand side of the highway (in the direction of Steele's Road, Glace Bay) in a large gravel pit which is undermined and has large cave-ins. The Emery seam outcrops (1977) along a man-made cliff of about 150 m length. This represents the periphery of a large gravel pit. The cliff walls are joint faces and readily collapse into the pit.

The road map was drawn from air photo A 21536-42, 1970, Cape Breton, Nova Scotia; National Air Photo Library Reproduction Centre, Energy, Mines and Resources, Ottawa, Ontario, Canada.

GEOGRAPHIC AND STRATIGRAPHIC LOCATIONS OF SPECIMENS F-1 TO F-682

F-1 to F-5, F-12 to F-16, F-32, F-36 to F-39, F-48-49:

Stratigraphic and geographic location unknown but assumed to be from the Morien series.

F-7, F-10, F-17 to F-20, F-22 to F-31, F-33 to F-35, F-40 to F-47:

The samples were collected from the Harbour seam at #12 Mine dump, Ling Street, New Waterford, N.S. It is assumed, as is usually the case, that the stone is cut from above the Harbour seam, i.e., the roof.

F-53 to F-78: Mc Aulay seam Fig. 9[1]

1 See note in Table 6 on spelling of Mc Aulay in Fig. 9.

Samples originated from debris and measured stratigraphic positions along the seashore (at low tide).
F-53 to F-57: stratigraphic position assumed to be above the seam, debris samples.
F-58 to F-62: see COAL SAMPLES, Appendix I.
F-63 to F-75: 3 to 5 cm above the seam.
F-76 to F-78: in the immediate vicinity of the seam but stratigraphic position unknown.

F-80 to F-102: Shoemaker seam Fig. 9

Positive identification of this seam on the ground is not certain, but by considering our detailed coastline traversing and the characteristic fossils *(L. eschweileriana)* found there, we believe that we sampled the Shoemaker seam.

All collected fossils come from the sea cliff and some are *in situ* samples. There are abundant lepidodendrid twigs at this locality.
F-80 to F-84: see COAL SAMPLES, Appendix I.
F-85 to F-87-2: 100 to 120 cm above the seam.
F-89 to F-91: 120 to 150 cm above the seam. This stratigraphic horizon is rich in lepidodendrid twigs.
F-92 to F-102: Exact stratigraphic position unknown but certainly more than 1 meter above the seam.

F-103 to F-113: Tracy seam, south of Wadden Cove, Cape Breton Island, Fig. 9

The old outcrop of Tracy in the sea cliff is marked by six posts which, in 1975, were 4 meters away from the present sea cliff (active erosion). Tracy was locally mined by the Hiawatha Mine and much old mining equipment is present, including coal carts. There are several old prospecting pits in this area.
F-103 to F-104: see COAL SAMPLES, Appendix I.
F-105 to F-113: about 3.5 meters above Tracy.

F-115 to F-175: Emery seam, Glace Bay, N.S.

This sample location is one of the many bootleg pits in this area. It was a very large operation (now defunct), and pillars were left standing to prevent cave-ins. We found iron tracks for coal carts about 35 meters from the entrance, and there were many large cross-cuts. Entrance into the pit is constricted by rock debris which poses danger. The exact location of the pit is omitted for this reason.

The stone is a grayish-shaly sandstone (called bluestone by local people), excellent for fossils, and tough. Large "pods" (stigmarian systems) are observable, some of which are 3 meters long with rhizophores clearly visible.

This pit is devoid of bats.

Samples were collected *in situ*, some from the mine dump.

This is a good location for various neuropterids, varieties of *Neuropteris scheuchzeri*, various alethopterids, and fern-like species such as *P. (Asterotheca) acadica*. *Odontopteris subcuneata* is rare.

F-115, F-116: see Figs. 10 and 11.

F-117: see COAL SAMPLES, Appendix I.

F-118 to F-175: a maximum of 60 cm above the seam.

F-176 to F-197: Phalen seam, Glace Bay, N.S., Fig. 13

The sample site is a gravel pit along Highway 255 in Glace Bay. The fossils were collected from the floor of this pit, which is stratigraphically above the Phalen seam. This location offers good chances for finding *Neuropteris tenuifolia*, *Lepidostrobophyllum* species and *Neuropteris (Mixoneura) ovata*.

F-176 to F-178: see COAL SAMPLES, Appendix I.

F-198 to F-203: unspecified locations in the Morien series

These samples are on permanent loan to the College of Cape Breton from Dr. G. MacLeod, Professor of Philosophy, College of Cape Breton.

F-204 to F-287: Stubbart seam, Prince Mine, Point Aconi, Cape Breton Island, Fig. 12

These fossils were collected underground in the Prince Mine and present *in situ* samples, except for the coal samples. Fossils are preserved in fine conditions ("proof-like"). Calamarians of *carinatus* type are abundant here.

F-205: see COAL SAMPLES, Appendix I.

F-208 to F-240: about 300 meters in the development slope, 5 to 15 cm above the seam.

F-243 to F-244: see COAL SAMPLES, Appendix I.

F-245 to F-287: collected at #1 West Development Slope, 5 to 25 cm above the seam.

L.O.-1-30; F-288 to F-382: Harbour seam at Lingan Mine, Lingan, Cape Breton Island

The samples were collected from the mine dump of the Lingan Mine. Rock is strictly from above the Harbour seam, up to 2 meters in the development areas (by personal observation).

It is of interest that slickensided shales of about 30 to 40 cm rest directly on the Harbour seam; this is generally interpreted as movements in a fault (no fossils have been observed).

The Lingan Mine dump is rich in diversified fern-like foliage and trees.

F-307, F-308, F-382: see COAL SAMPLES, Appendix I. See F-421 to F-425: fossils collected underground.

F-383 to F-391 (1976): Emery seam, Glace Bay, N.S.

See F-115 to F-175.

F-392 to F-405: Harbour seam, #26 Mine dump, Glace Bay, N.S.

These fossils are collected from the dump of #26 Mine; the rock is assumed to be from positions stratigraphically above the seam. According to the mine administration some rock may be undercuts.

F-406 to F-416: Bonar seam, Point Aconi, Cape Breton Island, Fig. 12

The samples were collected along the sea cliff at Point Aconi and were about 3 meters stratigraphically above the Upper Bonar seam.
F-406: see COAL SAMPLES, Appendix I.

F-417 to F-420: Stubbart seam, Point Aconi, Fig. 12

The samples were collected along the seashore representing a stratigraphic position of about 2 meters above the Stubbart seam (see arrows in Fig. 12).

F-421 to F-425: Harbour seam, underground in Lingan Mine

The samples were collected in the area of No. 4 Drive, #2 Slope.

F-427 to F-442: Harbour seam at Lingan Mine dump

See L.O.-1-30; F-288 to F-381.

F-443 to F-459-3: see Emery seam, Glace Bay, N.S.

The fossils were collected from 30 to 40 cm above the Emery seam.

F-460 to F-467: see Shoemaker seam

The samples were collected 150 cm stratigraphically above the seam *in situ* at the cliff.

F-468 to F-481: see Phalen seam, Glace Bay, N.S.

Sample numbers F-475 to F-479 area not in use.

F-482 to F-496-3: see Lingan Mine dump, Harbour seam

F-497: Joggins, N.S.; F-498 to F-502: unspecified coal measures, Cape Breton Island

On permanent loan to the College of Cape Breton from the Beaton Institute, College of Cape Breton, N.S. These specimens were collected well over 100 years ago by R. Brown (see *Linopteris obliqua).*

F-503 to F-507: see Phalen seam, Glace Bay, N.S.

F-508 to F-597: see Emery seam, Glace Bay, N.S.

F-598 to F-600: see COAL SAMPLES, Appendix I

F-601: Point Aconi, N.S., unspecified coal measure of the Morien series

Donated to the collection by David Tracz, Sydney, N.S. See also F-402-1 *(P. unitus).*

F-602 to F-622-1: Stubbart seam, Prince Mine, Point Aconi

F-602 to F-606: unknown location in the Prince Mine
F-607 to F-614: on #1 Slope, 100 feet north of #1 Level of Panel 2 West
F-615: unknown location in the Prince Mine
F-616 to F-618: 150 feet up #2 Slope from #1 Level of 2 West
F-619 to F-621: 100 feet up 1 Room of 2 West
F-622: top of #4 Slope, left side of entrance (facing downslope)
F-622-1: unknown location in the Prince Mine.

F-623 to F-629: Phalen seam in 1-B Colliery of No. 26 Mine, Glace Bay, N.S.

F-623: slant between new and empty roads, about 100 cm above Phalen
F-624 to F-626: on Travel Road at 605 feet 9 inches below sea level, about 30 cm above Phalen
F-627: in the vicinity of F-624, about 50 cm above Phalen
F-628 to F-629: on Travel Road at 603 feet 0 inches below sea level, about 10 cm above Phalen.

F-630 to F-632: see Tracy seam

F-633: see Shoemaker seam

F-634 to F-642-2: see Harbour seam, Lingan Mine dump

F-643: see Emery seam, Glace Bay, N.S.

F-644 to F-662: Stubbart seam, Prince Mine, Point Aconi

F-644: 1E belt level between #4 and #5 Slopes, about 150 cm above Stubbart
F-645 to F-646: top of #4 Slope, left side (facing down slope)
F-647 to F-653: unknown location in the Prince Mine
F-654 to F-655: intersection #4 Slope and 1E belt level
F-656 to F-662: unknown location in the Prince Mine.

F-670 to F-671: see Phalen seam, Glace Bay, N.S.

F-672 to F-674: see Stubbart seam, Prince Mine, Point Aconi

F-675 to F-682: Emery seam, at end of Steele's Road, Glace Bay, N.S., Fig. 13

The Emery seam outcrops 150 m along a cliff. Exact stratigraphic location of samples above the Emery is not known; the samples were collected from collapsed cliff walls.

APPENDIX I: COAL SAMPLES

Coal samples were collected for their association with minerals, for their own sake, and to investigate the replacement of plant matter by pyrite.

F-58 to F-62: Mc Aulay seam. Plate 148, Fig. 2

F-58: six cm from top of the seam; coal with masses of crystalline pyrite which are about six cm long and 1 to 2 cm thick.
F-59: pyrite in this sample is fine-grained and lenticular.
F-60: 6 to 8 cm from top of the coal seam; similar in grain size to F-59.
F-61, F-62: 10 to 15 cm from top of the seam; lenses of very fine-grained pyrite in the coal appear dark and indistinguishable from the coal.

In general the pyrite appears fresh and no oxidation alterations are noticeable. X-ray powder diffraction patterns (F-58) indicate the presence of minute traces of marcasite, the low-temperature polymorph of pyrite.

F-80 to F-84: Shoemaker seam

F-80, F-81: dirty coal with 1 to 4 mm lenses of fine-grained pyrite.
F-82: dirty coal with jarosite and fine-grained pyrite.
F-83: 21 cm above the seam; 5 to 9 cm plant fossils whose core is replaced by pyrite. The fossil cannot be determined. See Plate 147, Fig. 1.
F-84: 100 to 120 cm above Shoemaker seam; this specimen is similar to F-83 except for less replacement of vegetable matter by pyrite.

Jarosite, $K_2Fe_6(OH)_{12}(SO_4)_4$, is ubiquitous as lemon yellow tarnish on the dirty coal and plant fossils in this area, together with gypsum (X-ray powder diffraction analysis).

F-103 to F-104: Tracy seam

These samples were collected from the mine dump; no pyrite is visible in macroscopic examination.

F-117: Emery seam

Collected at 20 cm from roof of the Emery seam.

F-176 to F-178: Phalen seam

F-176: a large specimen of coal measuring 27 × 27 × 11 cm collected at 20 cm from the roof.
F-177: coal collected at 100 cm below the roof.
F-178: coal with lenses of fine-grained pyrite.

Assorted samples of coal were analyzed optically and by X-ray powder diffraction patterns. X-ray powder diffraction patterns revealed the presence of $FeSO_4.4H_20$ (rozenite, whitish material) and pyrite. Optical examination shows that pyrite is present in two forms (Plate 147, Fig. 2);

1) as nodules with apparently unstable core; this form is probably in the transition stage to marcasite; and
2) between the nodules elongated crystals were noticed perpendicular to the veins. This crystal morphology suggested marcasite, but closer investigation showed it to be pyrite.

No marcasite was detected in these samples.

F-205: Stubbart seam, Point Aconi, Prince Mine (underground samples)

Specimens of nodular pyrite which display a core section were collected from the center section of Stubbart seam.

F-243 to F-244: Stubbart seam; collected from the coal dump

F-243; coal with fine-grained pyrite.
F-244: coal with nodules of pyrite, i.e., a stigmarian is preserved by pyrite replacement. See Plate 148, Fig. 1.

F-307, F-308, F-382: Harbour seam, Lingan Mine dump

F-307: coal with gypsum veins?
F-308: coal with pyrite and chalcopyrite.
F-382: coal with oxidized pyrite. See Plate 149.

F-406: Bonar seam, Point Aconi

Coal is completely replaced by pyrite; the specimen is 2 cm thick and originated from the roof of the seam. See Plate 150.

F-598: Phalen seam

Coal with pyrite from #16 Dominion, donated by the Cape Breton Development Corporation, Coal Division.

F-599: Harbour seam, Princess Mine

From the last ton of coal mined at Princess, Jan. 1976. This mine was in production for over one hundred years.

F-600: coal from Lingan Mine, No. 3 East Wall, elevation 1150 feet, Harbour seam

Donated by the management of Lingan Mine.

Recent mineral collecting in the Phalen seam, 1-B Colliery at #26 Mine, 603-605 feet below sea level, reveals the presence of melanterite ($FeSO_4 \cdot 7H_2O$) in fair amounts. This is the second recorded locality in Canada, after the Manitoba site (Jambor and Traill, 1963). Melanterite occurs in the exposed parts of the Phalen seam, in mine workings over 60 years old, and it readily dehydrates at the surface to rozenite, $FeSO_4 \cdot 4H_2O$; see Phalen seam above, Zodrow and McCandlish (1978a; 1979).

The crystal habit of melanterite in the Phalen seam is fibrous, growing perpendicular to depositional planes of the coal faces, but also in massive nodules. For one possible origin of pyrite in coal see Casagrande and Siefert (1977).

APPENDIX II

Fern-like, fern-type, and other Carboniferous fossil plants in the collection of the Department of Geology, St. Francis Xavier University, Antigonish, Nova Scotia, Canada.

Collector and Determiner Abbreviations:

A	=	H.M. Ami, Geological Survey of Canada; see Zaslow (1975 p.147) for portraits.
AHF	=	A.H. Foord, Geological Survey of Canada; see Bell (1962, p. 63).
UNK	=	Unknown.
WAB	=	W.A. Bell, Geological Survey of Canada; see Zaslow (1975 p. 143) for a portrait.
WJW	=	W.J. Wilson, Geological Survey of Canada; see Zaslow (1975).

Macroplant Remains

Acitheca polymorpha (Brongniart)[1]
3817 (= sample number), 441 (= locality), at least for the specimens by Bell; see Bell (1938, p. 108-115).
Roof of the Bonar seam, Point Aconi, Morien series.
WAB

Alethopteris davreuxi (Brongniart) (for identity and synonymy see INDEX)
2507, 489
Roof of Phalen seam, Morien series.
WAB

Alethopteris scalariformis Bell
2606, 612
Roof of Mullins seam, Morien series.
UNK

Alethopteris valida Boulay
2278, 556
Roof of Phalen seam, Morien series.
WAB

Alloiopteris (Corynepteris) sternbergi (Ettingshausen)
9627, 1075
Cumberland series, Springhill, N.S.
UNK

1 We copied identification of species and its epithet, sample number, and sample location, if stated, from the accompanying label and reproduce it here; in the case of Bell's specimens we had his original hand-written labels. The meaning of the parentheses, as in "*A. polymorpha* (Brongniart)," is discussed on p. 22.

Annularia sphenophylloides Zenker (Bell did not use parentheses)
2220, 504
Shore of Cape Morien, Morien series.
WAB

Asolanus camptotaenia Wood
3440, 512
Roof of Phalen seam, Morien series.
UNK

Asterotheca crenulata (Brongniart)
3830, 532
Roof of Emery seam, Morien series.
See Bell (1938, p. 73) for relation between this species and *Asterotheca herdi* Bell.
WAB

Bellopteris corsini Radforth and Walton
= *Neuropteris (Bellopteris) corsini* Radforth and Walton
(Bell, 1962, p. 39).

Crossotheca boulayi Zeiller
*572
Pit in Mullin's seam, Morien series.
UNK

*In the case of a number by itself it is not known whether it is a locality or specimen number.

Eupecopteris dentata (Brongniart)
3818, 618
Roof of Emery seam, Morien series.
WAB

Lepidodendron pictoense Dawson
1171 (locality)
Pictou series, Clifton, N.B.
AHF, year 1880
See Bell (1962, p. 53) for arguments on the conspecificity of *L. lycopodioides* Sternberg, *L. lanceolatum* Lesquereux and this species.

Linopteris muensteri (Eichwald)
Sample number missing, 2880
Stellarton series.
UNK

L. muensteri (Eichwald)
2728
False Bay beach, Morien series.
WAB

Linopteris obliqua (Bunbury) Plate 20, Figs. 1, 2
2825, 447
Roof of Ormond seam, 1200 feet southwest of Sydney reservoir, Morien series.
WAB
The determination is disputable. It looks more like what Bell called *L. neuropteroides* Gutbier var. *major* H. Potonié, or *L. muensteri* (Eichwald) var. *dawsoni* Bell. It is difficult to be sure about this specimen (Dr. Arnold, *in litt.,* Sept. 1976). This specimen is similar to *L. obliqua* (Bunbury) Zeiller (Bell, 1966, p. 30, 28).

Linopteris obliqua (Bunbury) Zeiller Plate 19
7312, 897
Pit on Ormond seam, Mira Road, Morien series.
UNK (probably WAB, for location corresponds to Bell (1938, p. 112).
As in the previous case, the identity is disputable (Dr. Arnold, *in litt.,* Sept. 1976). Pinnule characteristics are similar to those of *Linopteris obliqua* (Bunbury) Zeiller as figured by Bell with locality 897 (1966, p. 34). This species, as well as *L. bunburii,* seems to be confined to this zone, equivalent to the Mullins seam (see Table 6).

Mariopteris latifolia (Brongniart)
4505, 2675
Welton Mine, N.B.
WJW, year 1910

Mariopteris latifolia (Brongniart) Plate 48
Locality 5074
Minto; Newcastle Mine dump, Pictou series, New Brunswick
UNK

Mariopteris nervosa (Brongniart)
3822, 618
Emery seam, Dominion #10 Colliery, Morien series.
WAB

Neuropteris (Bellopteris) corsini Radforth and Walton *(Bellopteris corsini* Radforth and Walton)
5074
Minto formation, Newcastle Mine, Pictou series, New Brunswick
UNK

Neuropteris (Mixoneura) obliqua (Brongniart)
Sample number missing
Gardners Creek, Cumberland series, New Brunswick
WJW, year 1908

Neuropteris scheuchzeri Hoffmann
3942, 477
Sydney Mines, Morien series.
WAB

Neuropteris smithsii Lesquereux
Sample number missing.
Riversdale series, Cumberland Co., N.S.
A, year 1898

N. smithsii Lesquereux
Locality 1392
Riversdale series, Inverness Co., Cape Breton Island.
UNK

Neuropteris tenuifolia (Schlotheim)
3781, 767
Roof of Mullin's seam, Morien series.
WAB

Oligocarpia brongniarti Stur
3819, 487
Between Emery and Phalen seams, Morien series.
WAB (refer to Table 8 for emending citation).

Pecopteris (Asterotheca) acadica Bell
3821, 715
Roof of Emery seam, Morien series.
WAB

Ptychocarpus unitus Brongniart
4465.C
Morien series.
UNK

Samaropsis cornuta (Dawson)
8033, 974
Pit south side of Cow Bay road, Morien series.
UNK

Senftenbergia plumosa Artis
Sample number missing.
Spicer Cove, Joggins, N.S., Cumberland series.
UNK

Sphenophyllum emarginatum Brongniart
1941, 614
Roof of Mullin's seam, Morien series.
WAB

Sphenophyllum oblongifolium Germar and Kaulfuss
2035, 428
Roof of Point Aconi seam, Morien series.
WAB

Sphenopteris obtusiloba Brongniart
4700, 990
Minto formation, Pictou series, New Brunswick
UNK

Sphenopteris spiniformis Kidston
3823, 482
Above Tracy seam north of False Bay beach, Morien series.
WAB

Sphenopteris sulcata Bell
1433
St. Mary's Point, Riversdale series, Albert Co., New Brunswick.
UNK

For further discussion of the conspecifics of fern-like foliage and other Carboniferous fossil plants of Nova Scotia, refer to Bell (1962).

Both Ami and Wilson were paleontologists for the G.S.C. who in 1908 worked in Antigonish County and in the Bay of Fundy region of New Brunswick, respectively. See G.S.C. Summary Report, 1908 for details (Zaslow, 1975, p. 298).

REFERENCES

Abbott, M. L. 1954. Revision of the Paleozoic fern genus *Oligocarpia*. Palaeontographica, (Series B) 96: 39-65.

Abbott, M. L. 1958. The American species of *Asterophyllites, Annularia,* and *Sphenophyllum*. Bull. Am. Paleontol. 38: 289-390.

Abbott, M. L. 1963. Lycopod fructifications from the Upper Freeport (No. 7) Coal in Southeastern Ohio. Palaeontographica, (Series B) 112:93-118.

Andrews, H. N. Jr. 1955. Index of generic names of fossil plants, 1820-1950. U.S. Geol. Survey Bull. 1013. 262 p.

Andrews, H. N. Jr. 1961. Studies in paleobotany. John Wiley and Sons, London. 487 p.

Archangelsky, S., and O. G. Arrondo. 1973. Paleophytologia Kurtziana, III. 10, The Permian flora of the Sierra de los Llanos, La Rioja, Argentina [in Spanish, English summary]. Ameghiniana, 10(3): 201-228.

Arnold, C. A. 1947. An introduction to paleobotany. McGraw-Hill, London. 433 p.

Arnold, C.A. 1948. The Mississippian flora. Jour. Geol. 56: 367-372.

Arnold, C. A. 1949. Fossil flora of the Michigan coal basin. Contr. Mich. Univ. Mus. Paleont., 7(9): 131-269, 34 pl.

Arnold, C. A. 1973. Fossil plants and continental drift. Birbal Sahni Inst. Paleobot., Lucknow, India. 11 p.

Banks, H. P. 1970. Evolution and plants of the past. Wadsworth Pub. Co., Belmont. 170 p.

Batenburg, L. H. 1977. The *Sphenophyllum* species in the Carboniferous flora of Holz (Westphalian D, Saar Basin, Germany). Rev. Palaeobot. Palynol. 24: 69-99.

Beck, C. B. 1970. Problems of generic delimitation in paleobotany. Proceedings of the North American Paleontological Convention, Chicago 1969. Part C. Allen Press, Lawrence, Kansas. p. 157-320.

Bell, W. A. 1938. Fossil flora of Sydney Coalfield, Nova Scotia. Geol. Surv. Can. Mem. 215. 334 p., 107 pl.

Bell, W. A. 1940. The Pictou Coalfield, Nova Scotia. Geol. Surv. Can. Mem. 225. 160 p., 10 pl.

Bell, W. A. 1944. Carboniferous rocks and fossil flora of Northern Nova Scotia. Geol. Surv. Can. Mem. 238. 276 p., 79 pl.

Bell, W. A. 1962. Flora of Pennsylvanian Pictou Group of New Brunswick. Geol. Surv. Can. Bull. 87. 71 p., 56 pl.

Bell, W. A. 1962a. Catalogue of types and figured specimens of fossil plants in the Geological Survey of Canada Collections. Geol. Surv. Can. 154 p.

Bell, W. A. 1966. Carboniferous plants of Eastern Canada. Geol. Surv. Can. Paper 66-11. 76 p., 36 pl.

Blackadar, R. G., H. Dumych and P. J. Griffin. 1975. Guide to authors [revised edition]. Geol. Surv. Can. Misc. Report 16. 199 p.

Boersma, M. 1978. A survey of the fossil flora of the Illinger Floezzone (Heusweiler Schichten, Lower Stephanian, Saar, German Federal Republic). Rev. Palaeobot. Palynol. 26:41-92.

Brack-Hanes, S.D., and J. C. Vaughan, 1978. Evidence of Paleozoic chromosomes from lycopod microgametophytes. Science, 200: 1383-85.

Brongniart, A. 1822. Sur la classification et la distribution des végétaux fossiles en général, et sur ceux des terrains de sédiment supérieur en particulier. Mém. Mus. d'Hist. nat. [vol. 8], Paris. p. 203-348.

Brongniart, A. 1827-1830. Histoire des végétaux fossiles ou recherches botaniques et géologiques sur les végétaux renfermes dans les diverses couches du globe. 2 vols. and atlas. Paris.

Brown, R. 1871. The coal fields and coal trade of the Island of Cape Breton. Sampson, Low, Marston, Low & Searl, London. 166 p.

Bunbury, C. J. F. 1847. On fossil plants from the coal formation of Cape Breton. Journ. Geol. Soc. London, 3: 423-438, 4 pl.

Burbridge, P., and C. J. Felix. 1976. Stratigraphic and morphologic development in the spore genus *Spencerisporites*. Geoscience and Man, 15: 87-94.

Casagrande, D., and K. Siefert. 1977. Origins of sulfur in coal: importance of the ester sulfate content of peat. Science, 195: 675-676.

Chandra, S. 1974. Ferns and fern-like plants from the Lower Gondwana. Birbal Sahni Inst. Paleobot., Lucknow, India. p. 62-72.

Corsin, P. 1951. Pécoptéridées. *In* Études des Gîtes Minéraux. Bassin Houiller de la Sarre et de la Lorraine. I. Flore fossile, 4th fascicle, p. 177-370, pls. 108-199. Service des Topographies Sousterraines, Loos-Nord.

Crookall, R. 1929. Coal measure plants. E. Arnold, London. 70 p.

Crookall, R. 1955-1970. Fossil plants of the Carboniferous rocks of Great Britain. Geol. Surv. Great Britain, vol. 4, pts. 1-6.

Daber, R. 1955. Pflanzengeographische Besonderheiten der Karbonflora des Zwickau-Lugauer Steinkohlenreviers. Geologie (Berlin), Suppl. 13. 44 p., 25 pl.

Darrah, W. C. 1960. Principles of paleobotany [2nd edition]. Ronald Press, New York. 295 p.

Dawson, J. W. 1878. Acadian geology [3rd edition]. Macmillan, London. 694 p.

Dijkstra, S. J. [ed.] Fossilium Catalogus. Section II. Plantae Pars 48-87. (Note: no date appears because this periodical is published continuously updating plant fossil names; see also Jongmans, W.)

Doubinger, J. 1956. Contribution à l'étude des flores autuno-stephaniennes. Mém. Soc. Géol. France, new series, Vol. XXXV(75) Paris. 180 p., 16 pls.

Fefilova, L. A. 1968. European and American ferns in Permian deposits of the Pechora region of the Urals [English translation]. Akad. Nauk SSSR, Dokl. 183(3): 680-682.

Fefilova, L. A. 1973. Permian ferns of the northern part of the Ural foredeep [in Russian]. Izd. Nauka, Leningrad. 190 p.

Fraas, E. 1976. Der Petrefaktensammler [4th edition]. Franckh'sche Verlagshandlung, Stuttgart. 312 p., 72 pl.

Grauvogel-Stamm, L., and J. Doubinger. 1975. Two fertile ferns from the Stephanian of the Central Massif, France. Geobios, 8(6): 409-421.

Gregory, D. [compiler]. 1975. Bibliography of the geology of Nova Scotia. Nova Scotia Department of Mines, Halifax. 237 p.

Gothan, W. 1953. Die Steinkohlenflora der westlichen paralischen Steinkohlenreviere Deutschlands. Geol. Jahrbuch, vol. 10 (suppl.). 83 p., 44 pl.

Gothan, W., and H. Weyland. 1964. Lehrbuch der Palaeobotanik [2nd edition]. Akademie-Verlag, Berlin. 594 p.

Gothan, W., and H. Weyland. 1973. Lehrbuch der Palaeobotanik [3rd edition]. BLV Verlagsgesellschaft, Munich. 677 p.

Havlena, V. 1953. The Neuropterides of the Carboniferous and Permian of Bohemia [English translation of p. 131-168]. Czech. Akad. Geol. Ustavu, 16: 1-168, 8 pls.

Herbst, R. 1972. *Gleichenites potrerillensis* new species from the Middle Triassic of Mendoza, Argentina, with comments on fossil Gleichenaceae of Argentina. Ameghiniana, 9(1): 17-22.

Jambor, J. L., and R. J. Traill. 1963. On rozenite and siderotil. Canad. Mineralogist, 7(5): 751-763.

Jennings, J. R., and D. A. Eggert. 1972. *Seftenbergia* is not a schizaeaceous fern. Am. Journ. Bot., 59(6): 676.

Jongmans, W., [ed.] Fossilium Catalogus. Section II. Plantae Pars 1-48. (Note: no date appears because this is a continuously published periodical updating plant fossil names: see also Dijkstra, S.J.)

Khramova, S. N., and V. V. Pavlov. 1971. Some ferns from the Upper Triassic deposits in the Timan-Pechora area [in Russian, English summary]. Paleontol. Abh. (Series B), 10: 71-77.

Kidston, R. 1923-25. Fossil plants of the Carboniferous rocks of Great Britain. Geol. Surv. Great Britain, Paleontology, vol. 2 (parts 1-6), 681 p.

Kodak Publication No. N-17. 1974. Kodak infrared films. Eastman Kodak Co., Rochester, New York. 16 p.

Kummel, B. 1970. History of the earth. Freeman, San Francisco. 707 p.

Lindley, J., and W. Hutton. 1831-37. The fossil flora of Great Britain; or figures and descriptions of the vegetable remains in a fossil state in this country. 3 vols. Ridgeway, London.

LeRoux, S. F. 1976. On some northern elements in the Lower Gondwana flora of Vereeniging Transvaal, Africa. Palaeontologia Africana, 19: 59-65.

Linneai, C. 1740. Systema naturae in quo naturae regna tri, secundum. classes, ordines, genera, species. Kiesewetter, Stockholmiae.

Mamay, S. H. 1950. Some American Carboniferous fern fructifications. Ann. Missouri Bot. Gard., 37: 409-476.

Matthews, R. K. 1974. Dynamic stratigraphy. Prentice Hall, Englewood Cliffs. 370 p.

Milligan, G. C. 1977. The changing earth: An introduction to geology. McGraw-Hill Ryerson, Toronto. 706 p.

Mitchison, G. J. 1977. Phyllotaxis and the Fibonacci series. Science, 196: 270-275.

Moore, R. C. *et al.* [editors]. 1944. Correlation of Pennsylvanian formations of North America. G.S.A. Bull. 55: 657-706.

Novik, E. 1952. Flores du houllier de la partie Européene de l'U.R.S.S. Acad. Sc. U.R.S.S. Inst. Paleontol. Series Paleontol. de l'U.R.S.S. Nulle sér. Tome 1. 468 p.

Orlov, Yu. A. [editor]. 1962. Fundamentals of paleonotology [Translated from Russian]. Israel Program for Scientific Translations, Jerusalem. 482. p.

Rudwick, M. J. S. 1976. The meaning of fossils [2nd revised edition]. Neale Watson, New York. 287 p.

Robb, C. 1876. Report on explorations and surveys in Cape Breton, N.S. Can. Geol. Surv. Rept. Prog. 1874-75, p. 166-266.

Schlotheim, von, E. F. 1820. Die Petrefactenkunde auf ihrem jetzigen Standpunkte durch die Beschreibung seiner Sammlung versteinerter u. fossiler Ueberreste des Thier u. Pflanzenreichs der Vorwelt. Hofbuchdruckerey, Gotha. 437 p., XV pl.

Sitar, V., and J. Vozar. 1973. Die ersten Makrofloren-Funde in dem Karbon der Choc-Einheit in der Niederen Tatra (Westkarpaten). Geol. Zb. (Slov. Akad. Vied), 24(2): 441-448.

Stafleu, F. A., *et al.* [editors]. 1972. International code of botanical nomenclature. A. Oosthoek's Uitgeversmaatschappij N.V., Utrecht. 426 p.

Stewart, W. N., and T. Delevoryas. 1956. The medullosan pteridosperms. Bot. Rev. 22: 45-80.

Stockmans, F. 1933. Les neuroptéridées des bassins houillers Belges. Mém. Mus. roy. d'Hist. Nat., Brussels. No. 57, 61 p., 16 pl.

Stoll, N. R., *et al.* [editors]. 1961. International code of Zoological Nomenclature adopted by the XV International Congress of Zoology. International Trust for Zoological Nomenclature, London. 176 p.

Storch, D. 1966. Die Arten der Gattung *Sphenophyllum* Brongniart im Zwickau-Lugau-Oelnitzer Steinkohlenrevier. Palaeontol. Abhandl., Abt. B. 2(2): 195-326.

Stur, D. 1877. Beitraege zur Kenntniss der Flora der Vorwelt: Die Culm-Flora. 2. Die Culm-Flora der Ostrauer und Waldenburger Schichten. K.-k. geol. Reichsanstallt, Abhandl., Vienna. 8: 107-472, 18-44 pl.

Wagner, R. H. 1962. A brief review of the stratigraphy and floral succession of the Carboniferous in NW. Spain. Compte Rendu, Tome III. Quatrième Congrès pour l'avancement des études de stratigraphie et de géologie Carbonifère. p. 753-761.

Wegener, A. 1929. The origin of continents and oceans. [Trans. from German by J. Biram]. Dover Publications, New York. 246 p.

Weiss, C. E. 1869. Fossile Flora der juengsten Steinkohlenformation und des Rothliegenden im Saar-Rhein Gebiete. Teil 1. K. Akad. der Wiss. Berlin. 100 p., 12 pl.

Wendt, H. 1965. Ich suchte Adam. Rororo, Rastatt. 503 p.

White, D. 1943. Lower Pennsylvanian species of *Mariopteris, Eremopteris, Diplothmema,* and *Aneimites* from the Appalachian Region. U.S. Dept. Int., Washington. Prof. Paper 197-C, 85-140 p., 32 pl.

Zaslow, M. 1975. Reading the rocks. Macmillan, Toronto. 599 p.

Zodrow, E. L., and K. McCandlish. 1978. Distribution of *Linopteris obliqua* in the Sydney Coalfield of Cape Breton, Nova Scotia. Palaeontographica, (Series B) 168: 1-16.

Zodrow, E. L. and K. McCandlish. 1978a. Hydrated sulfates in the Sydney Coalfield, Cape Breton, Nova Scotia. Canad. Mineral. 16:17-22.

Zodrow, E. L. and K. McCandlish. 1979. Hydrated sulfates in the Sydney Coalfield of Cape Breton, Nova Scotia. II. Pyrite and its alteration products. Canad. Mineral. In press.

Zodrow, E. L. and K. McCandlish. 1979a. Note on *Trigonocarpus* type seed attached to *Neuropteris (Mixoneura) flexuosa* from the Sydney Coalfield, Cape Breton, Nova Scotia, Canada. In review.

Zodrow, E.L. 1979b. Fossils and sulfates of Sydney Coalfield, Cape Breton, N.S. In press.

Zodrow, E.L. 1979c. The aluminocopiapites of Sydney Coalfield. [unpublished].

Plates

Photography by
K. McCandlish and
E. L. Zodrow

Plate 1

Fig. 1 / *Alethopteris davreuxi* (Brongniart) (F-337). Harbour seam, Lingan Mine (p. 27); shown are detail of venation, mode of attachment of pinnules to rachis (decurrent) and shape of pinnules.

Fig. 2 / *Alethopteris davreuxi* (Brongniart) (F-212). Stubbart seam, Prince Mine; pinna with indicated apical parts.

Fig. 3 / *Alethopteris davreuxi* (Brongniart) (F-522). Emery seam, Glace Bay; attached pinnae.

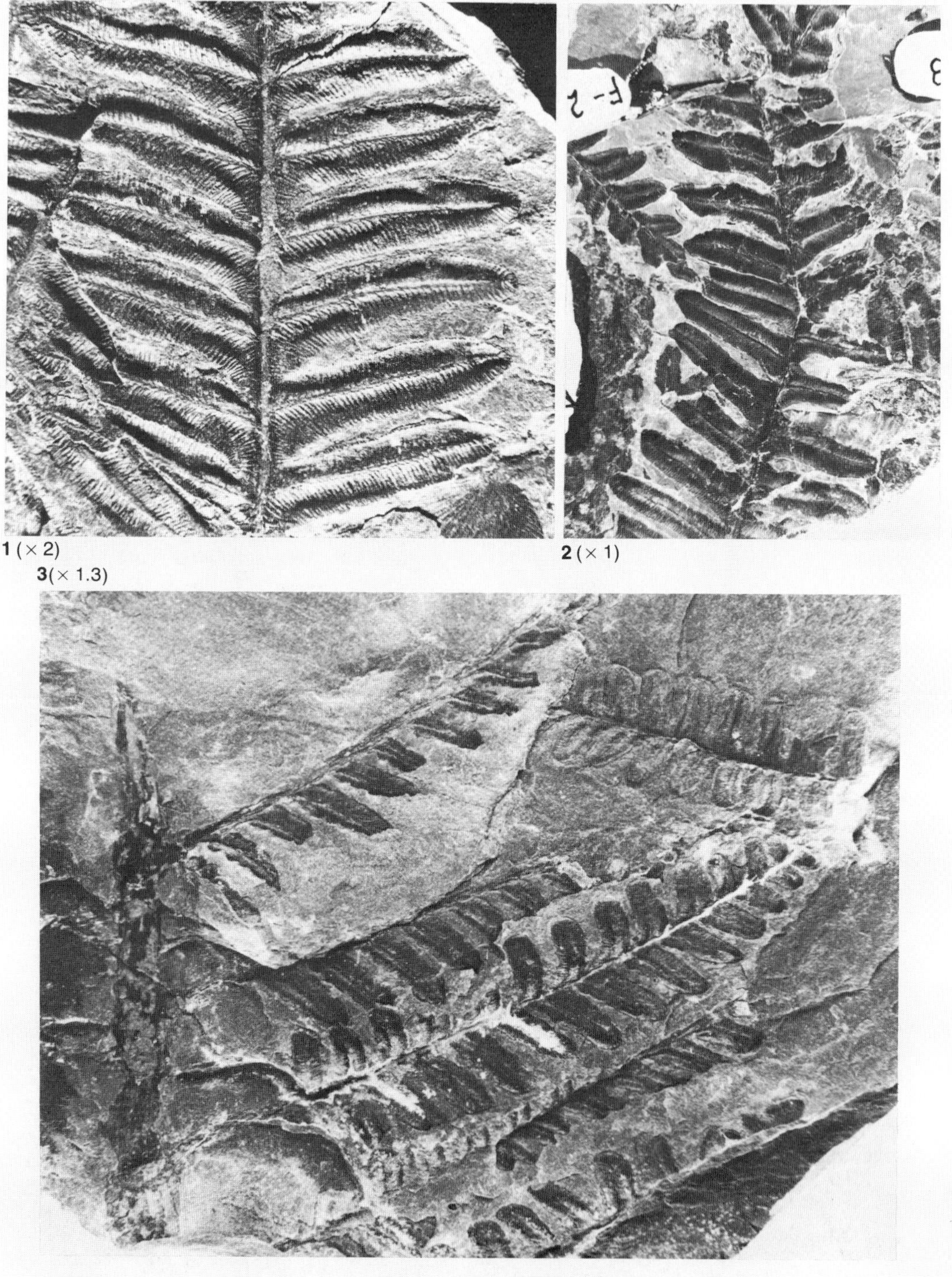

1 (× 2) **2** (× 1) **3**(× 1.3)

1 (× 2.7)

2 (× 1.3)

3 (× 2)

4 (× 2.6)

Plate 2

Fig. 1 / *Alethopteris davreuxi* (Brongniart) (F-361). Harbour seam, Lingan Mine (p. 27); fragment of a pinna.

Fig. 2 / *Alethopteris davreuxi* (Brongniart) (967G16.15). Harbour seam, Florence Mine; apical pinna.

Fig. 3 / *Alethopteris davreuxi* (Brongniart) (F-329). Harbour seam, Lingan Mine; apical part of a pinna.

Fig. 4 / *Alethopteris davreuxi* (Brongniart) (F-4). Assumed to be from the Sydney Coalfield (= Morien series); detail of an upper part of a pinna.

Plate 3

Fig. 1 / *Alethopteris decurrens* (Artis) (976GF22.1). Assumed to be from Nova Scotia (p. 28).

Fig. 2 / Detail of (976GF22.1). Showing venation pattern in lower pinnules.

Fig. 3 / *Alethopteris davreuxi* (Brongniart) (F-360). Harbour seam, Lingan Mine (p. 27); central part of a pinna.

2 (× 12)

3 (× 1.3)

1 (× 3)

1 (× 1)

2 (× 1)

3 (× 2)

Plate 4

Fig. 1 / *Alethopteris grandini* Brongniart (967G123.18). Morien series, specific location unknown (p. 28); a partial pinna.

Fig. 2 / *Alethopteris scalariformis* Bell (967G10.73). Sample location is unknown, may be from Cape Breton Island (p. 29).

Fig. 3 / *Alethopteris lonchitica* (Schlotheim) (967G10.72) Specific location in the Morien series is unknown (p. 29).

Plate 5

Fig. 1 / *Alethopteris scalariformis* Bell (967G123.22). Morien series, Cape Breton Island (p. 29); partial pinna.

1 (× 2.5)

1 (× 1)

2 (× 1.3)

3 (× 2)

Plate 6

Fig. 1 / *Alethopteris serli* (Brongniart) (F-594). Emery seam in Glace Bay, (p. 29); a large partial pinna.

Fig. 2 / *Alethopteris serli* (Brongniart) (F-149). Same location as in Fig. 1; ultimate part of a pinna.

Fig. 3 / *Alethopteris serli* (Brongniart) (F-388). Same location as in Fig. 1; partial pinna.

Plate 7

Fig. 1 / *Alethopteris serli* (Brongniart) (F-568). Emery seam, Glace Bay (p. 29); terminal part of a frond.

Fig. 2 / *Alethopteris serli* (Brongniart) (F-144). Same location as in Fig. 1; center part of a pinna.

Fig. 3 / *Alethopteris valida* Boulay (F-570). Same location (p. 30) as in Fig. 1; showing detail of attachment of unusually large pinnules to the rachis (lower border of the photograph).

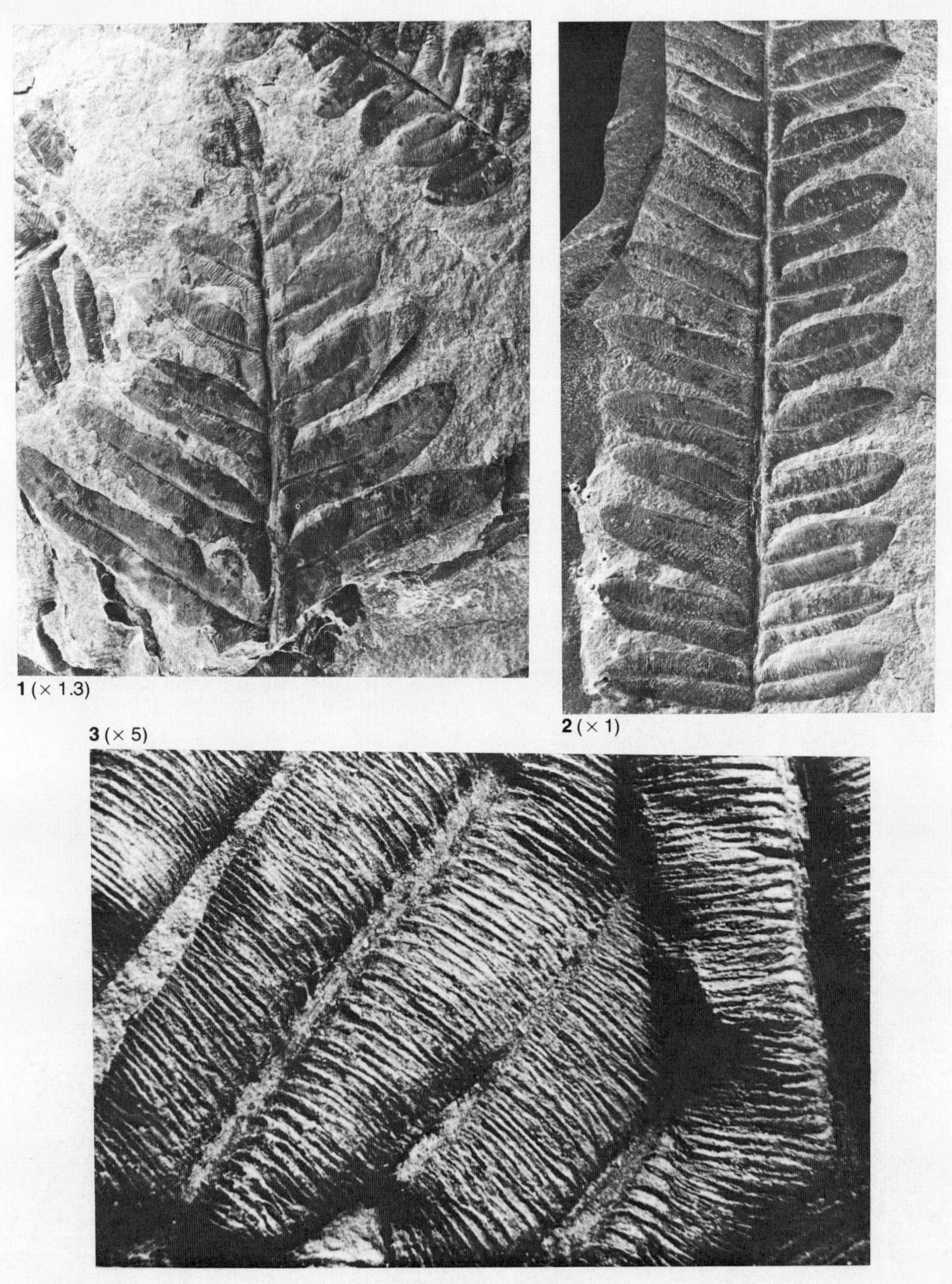

Plate 8

Fig. 1 / *Alethopteris valida* Boulay (F-586) verso. Emery seam, Glace Bay (p. 30); fragment of a pinna with rather large pinnules; compare with Plate 7, Fig. 3. Infrared reflection image.

Fig. 2 / *Alethopteris serli* (Brongniart) (F-154). Emery seam, Glace Bay (p. 29).

Fig. 3 / *Alethopteris serli* (Brongniart) (F-143). Emery seam, Glace Bay.

Fig. 4 / *Alethopteris serli* (Brongniart) (F-389). Emery seam, Glace Bay.

Plate 9

Fig. 1 / *Alethopteris serli* (Brongniart) (F-150). Emery seam, Glace Bay (p. 29).

Fig. 2 / *Alethopteris valida* Boulay, on display at the Nova Scotia Museum, Halifax. Originally from the Florence Mine, Cape Breton Island (p. 30).

1 (× 1)

2 (× 1)

(3 × 2)

Plate 9 continued

Fig. 3 / *Alethopteris serli* (Brongniart) (F-618). Stubbart seam, Prince Mine; note four apical parts of pinnae (all fossil plants in this photograph are the same species). Compare with other specimens of *A. serli* (p. 29).

Plate 10

Fig. 1 / *Alethopteris valida* Boulay (F-359). Harbour seam, Lingan Mine (p. 30); partial pinna.

Fig. 2 / *Alethopteris valida* Boulay (F-456). Emery seam, Glace Bay.

Fig. 3 / *Alethopteris valida* Boulay (F-455). Same location as in Fig. 2; this could be the ultimate pinna of the *Validopteris integra* Gothan type.

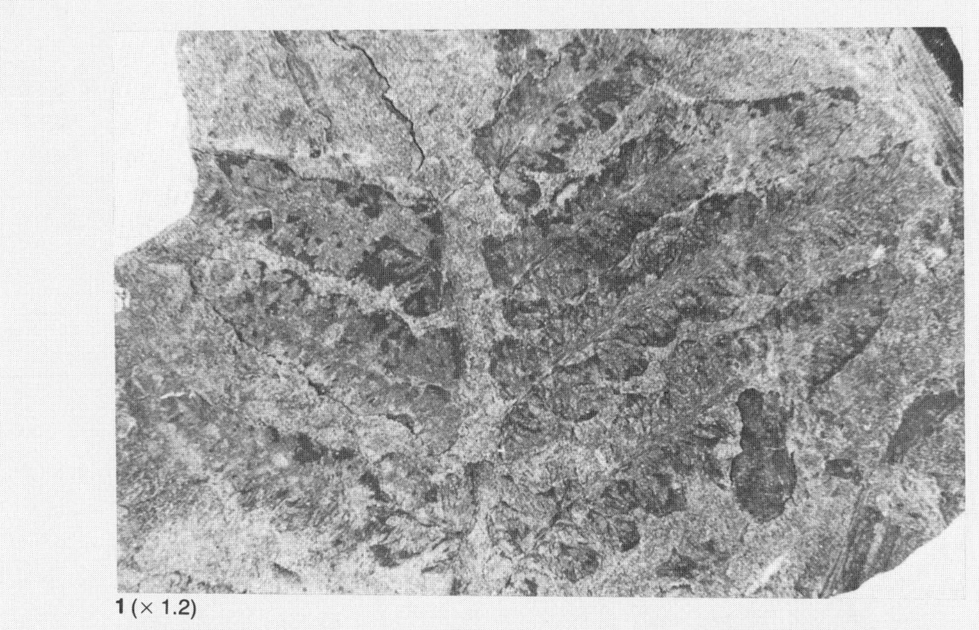

1 (× 1.2)

2 (× 6)

Plate 11

Fig. 1 / *Lonchopteris eschweileriana* Andrae (F-94). Shoemaker seam (p. 31); a partial frond.

Fig. 2 / *Lonchopteris eschweileriana* Andrae (F-460). Shoemaker seam; detail of a partial frond. See the following plate for venation detail of other examples and note open-meshed venation structure in all specimens.

Plate 12

Fig. 1 / *Lonchopteris eschweileriana* Andrae (F-460b). Shoemaker seam; note converging vascular strands near margin of pinnule (p. 31).

Fig. 2 / *Lonchopteris eschweileriana* Andrae (F-390). Emery seam, Glace Bay.

Fig. 3 / *Lonchopteris eschweileriana* Andrae (F-461). Shoemaker seam; portion of margin of lobate pinnules. Note the open-meshed arrangement of veins; see also Fig. 1.

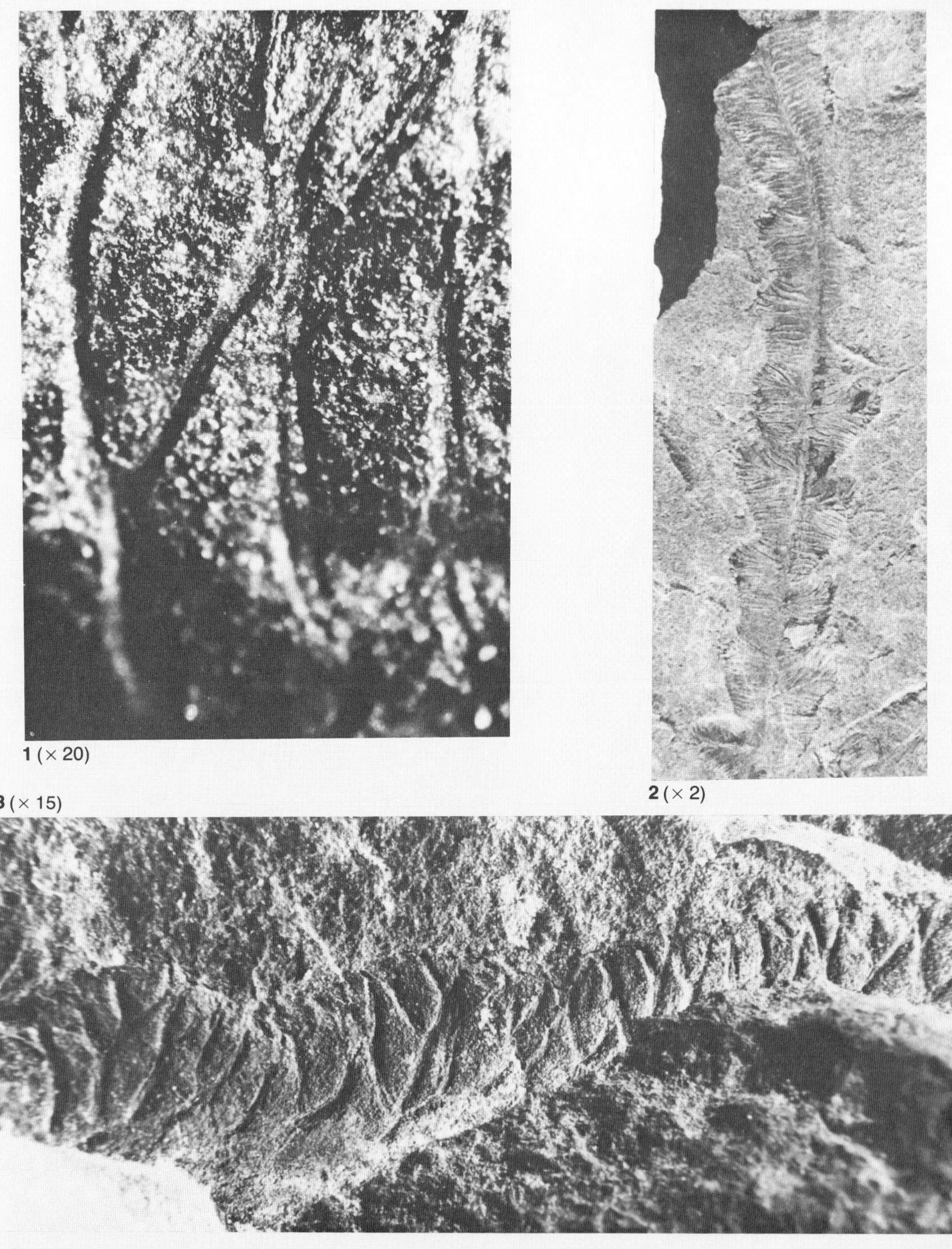

1 (× 20)

2 (× 2)

3 (× 15)

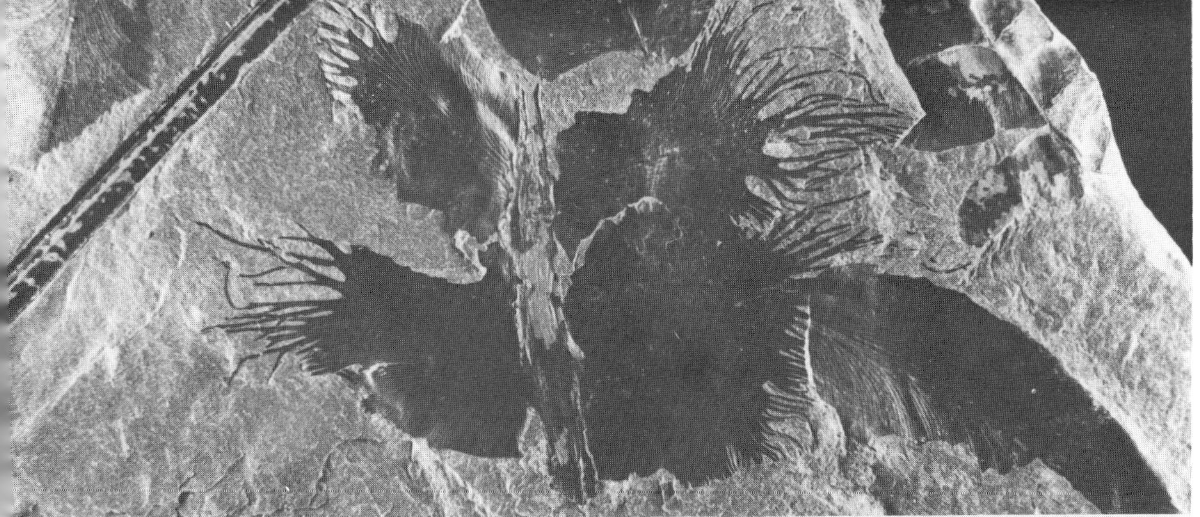

1 (× 1.5)

2 (× 2)

3 (× 3)

Plate 13

Fig. 1 / *Cyclopteris* cf. *fimbriata* Lesquereux (F-443). Emery seam, Glace Bay (p. 32); fragment of a structure.

Fig. 2/ *Cyclopteris* cf. *fimbriata* Lesquereux (F-378). Harbour seam, Lingan Mine; terminal part of the plant structure.

Fig. 3 / *Linopteris muensteri* (Eichwald) (F-334). Harbour seam, Lingan Mine (p. 33); detached pinnule.

Plate 14

Fig. 1 / *Cyclopteris* species (F-430). Harbour seam, Lingan Mine (p. 32); detached leaflet.

Fig. 2 / *Linopteris muensteri* (Eichwald), on block of 5074 (Appendix II, St. F. X. Collection p. 109); detail of a pinnule belonging to a partial pinna. This type of open-meshed veining is distinctive and can hardly be confused with *L. obliqua* (Bunbury) nor with varieties and species of the genus *Linopteris* Presl; some *Neuropteris* species display some anastomose form. See Plate 48 for associated *Mariopteris latifolia*.

1 (× 2)

2 (× 10)

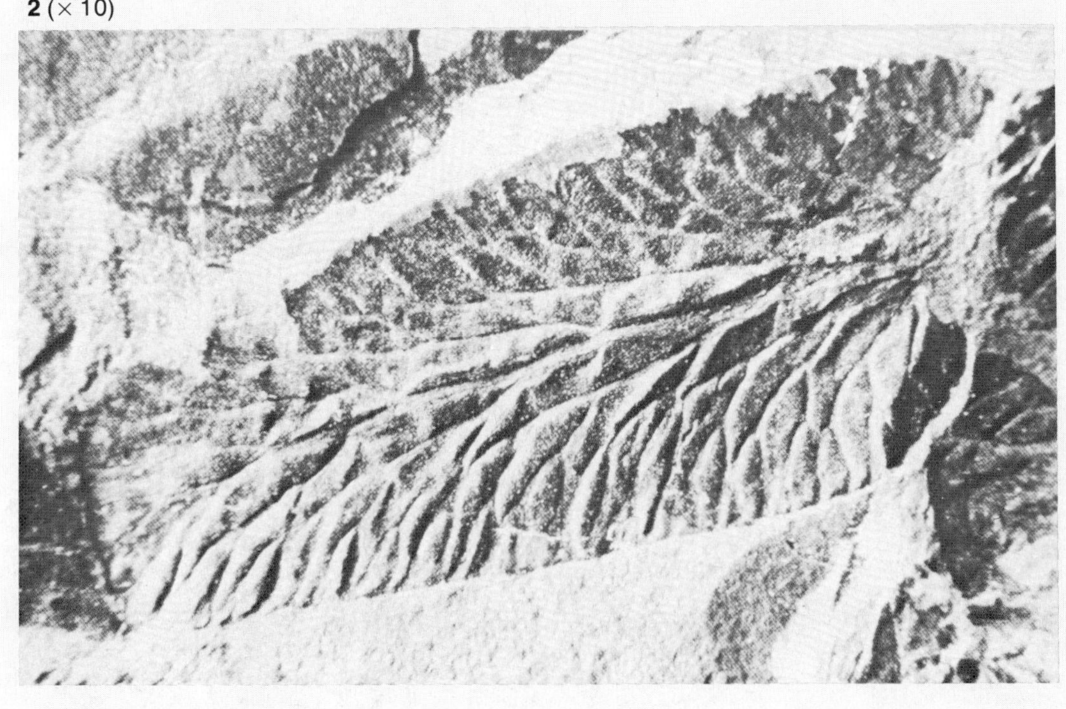

1 (× 5)

2 (× 2)

Plate 15

Fig. 1 / *Linopteris neuropteroides* (Gutbier) (F-110). Tracy seam (p. 33); entire specimen is shown.

Fig. 2 / *Linopteris obliqua* (Bunbury) (F-263). Stubbart seam, Point Aconi, Prince Mine (p. 33); detached pinnules.

Plate 16

Fig. 1 / *Linopteris obliqua* (Bunbury) (F-229). Stubbart seam, Point Aconi, Prince Mine (p. 33); residual carbon (lower of part of the photograph) is visible only under infrared reflection image; see Plate 17, Fig. 3.

Fig. 2 / Detail of (F-229) showing venation pattern; note densely distributed punctae (in elongated hexagons) which are interpreted as bases of hairs. (Bell, 1938, p. 64).

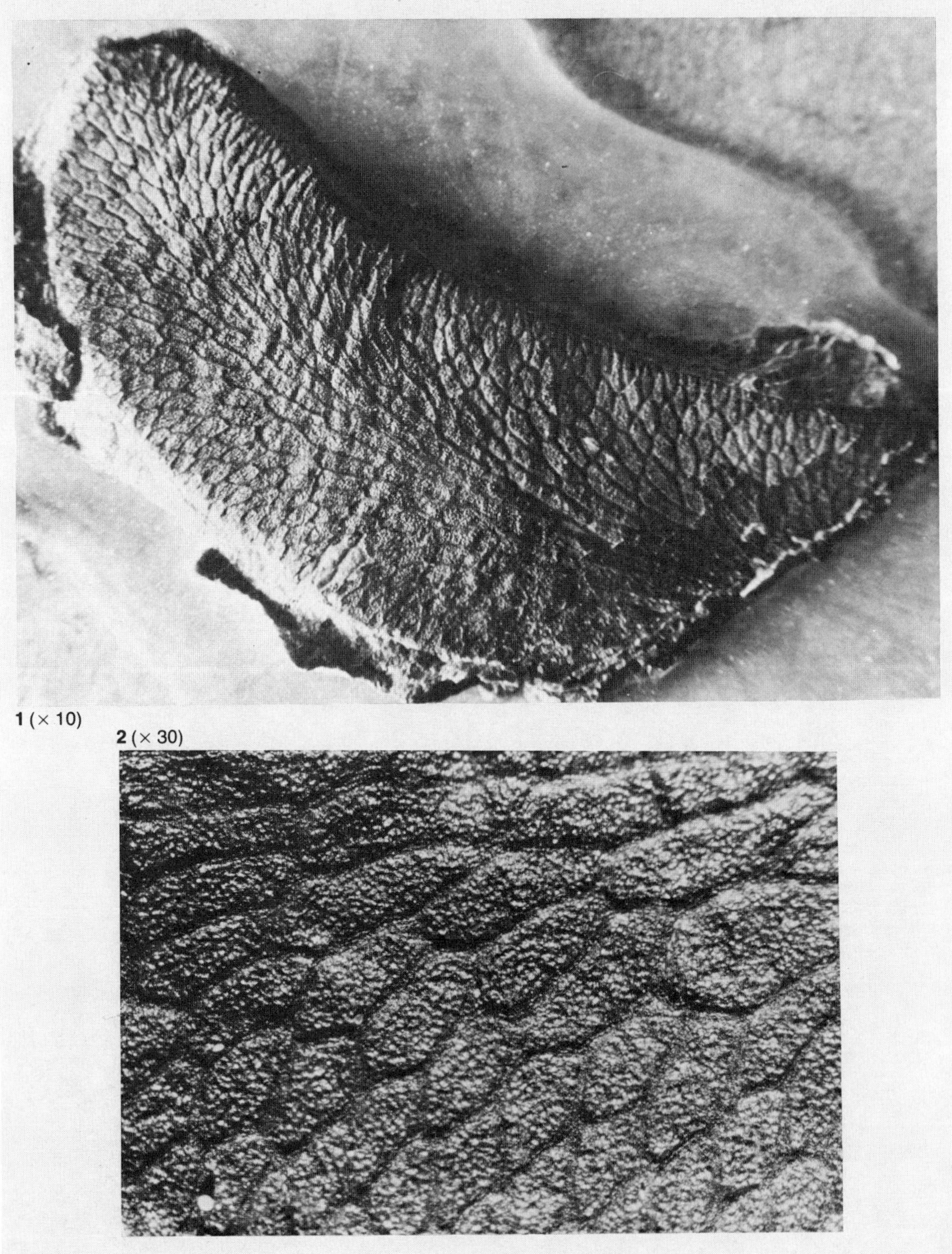

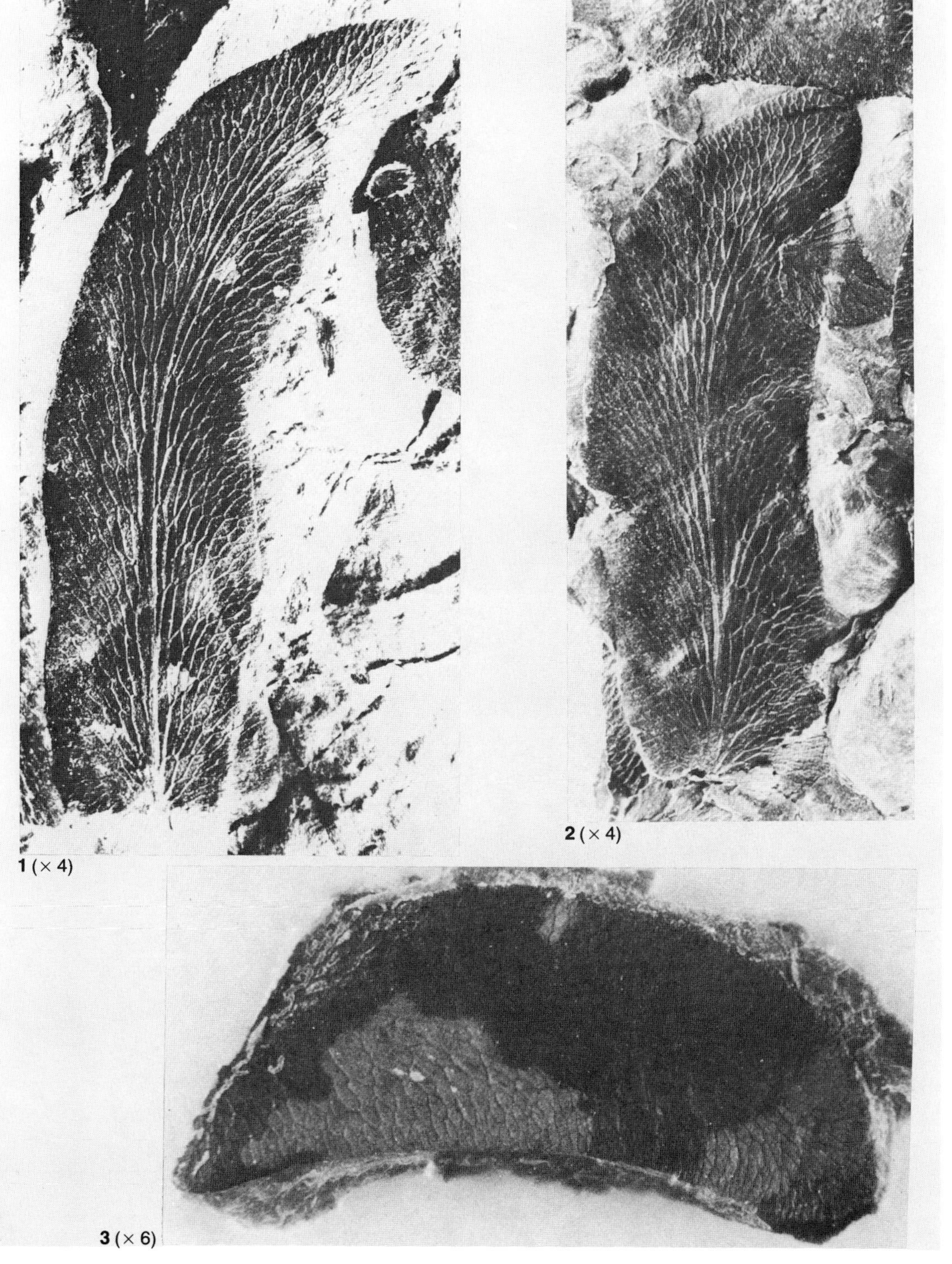

Plate 17

Fig. 1 / *Linopteris obliqua* (Bunbury) (F-644). Stubbart seam, Point Aconi, Prince Mine, showing typical venation pattern of elongated hexagons, steeply ascending, then tending towards orthogonality near margins; note the stout midrib which is traceable for about three-quarters of the pinnule length (p. 33).

Fig. 2 / *Linopteris obliqua* (Bunbury) (F-644). Same location as above, but a different specimen of the species on this block.

Fig. 3 / *Linopteris obliqua* (Bunbury) (F-229). Same location as in Fig. 1. Residual organic carbon is well contrasted as black area, but venation pattern is obscured by infrared reflection image; compare to Plate 16, Fig. 1.

Plate 18

Fig. 1 / *Linopteris obliqua* (Bunbury) (L.O.-25). Harbour seam, Lingan Mine (p. 33); detached pinnules. The arrow identifies the pinnule enlarged in Figs. 2 and 3 below.

Fig. 2 / Enlarged pinnule (L.O.-25) of specimen marked by arrow in Fig. 1. Infrared reflection image; note that venation detail is lost, but residual carbon is visible in the lower left portion of the base.

Fig. 3 / Enlarged pinnule (L.O.-25) of specimen marked by arrow in Fig. 1. Refer to Fig. 2 and note detail of venation in this figure.

1 (× 14)

Plate 19

Fig. 1 / *Linopteris obliqua* (Bunbury) (7312,897). Ormond seam; detail of venation. Note the slender meshing of veins and nondescript midrib; center area of pinnule is shown. See Appendix II, p. 109.

Plate 20

Fig. 1 / *Linopteris obliqua* (Bunbury) (2825, 447). Ormond seam; upper part of the pinnule. See Appendix II, p. 109.

Fig. 2 / Infrared reflection image of above specimen; see INTRODUCTION, p. 20, for comments on contrast and resolution.

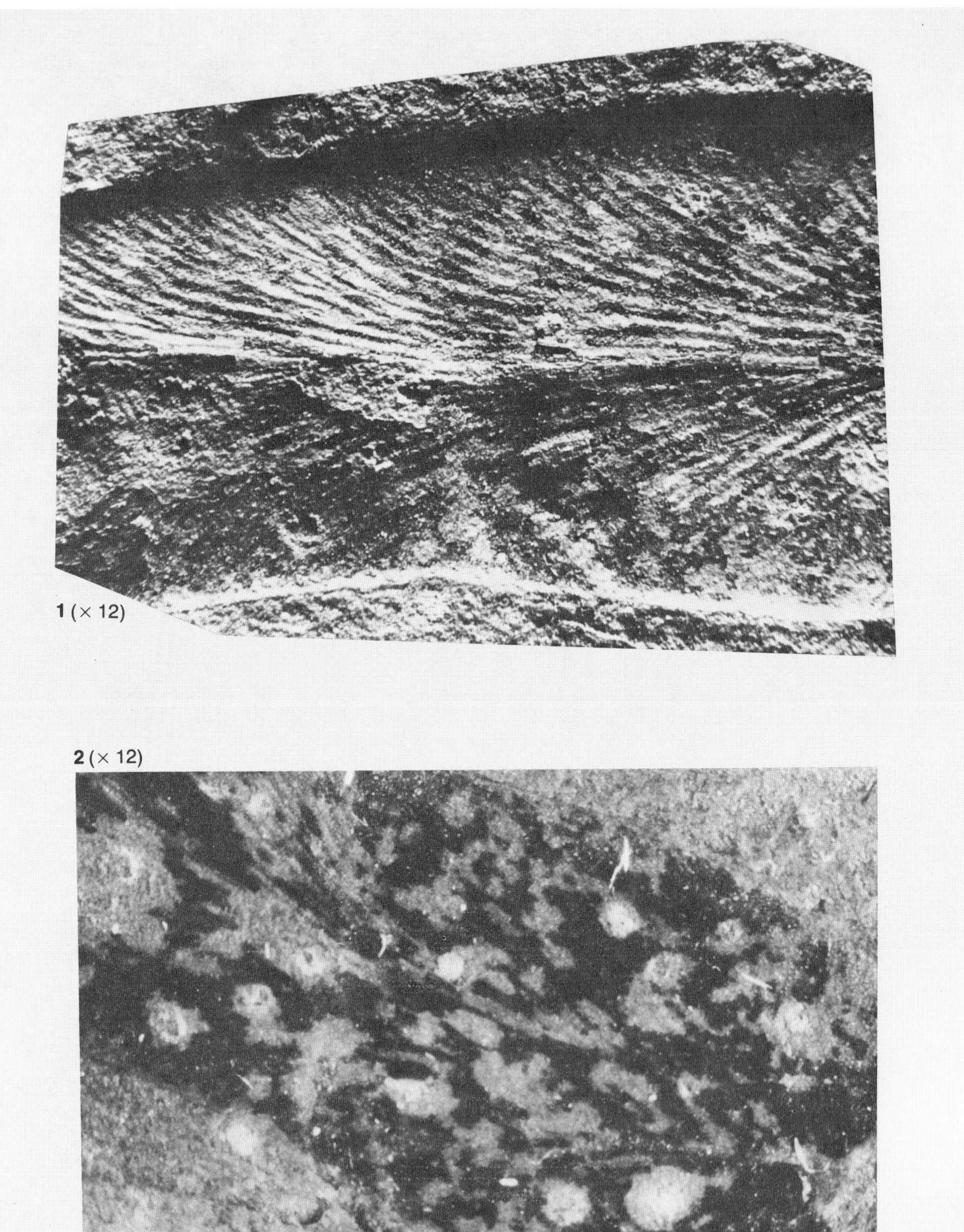

Plate 21

Fig. 1 / *Linopteris obliqua* (Bunbury) (L.O.-27). Harbour seam, Lingan Mine (p. 33); some partially attached pinnules.

Fig. 2 / *Linopteris obliqua* (Bunbury) var. *bunburii* Bell (= *L. bunburii*), paratype GSC 7052; Bell (1938, Pl. 60, Fig. 5); roof of Emery seam (p. 34). By permission Geological Survey of Canada. This specimen may be compared to Bell's Fig. 5 (Bell, 1966, p. 46, Pl. 22).
See also Fig. 4 and Plate 22, Fig. 1.

Fig. 3 / *Linopteris bunburii* Bell (F-285-3). Stubbart seam, Point Aconi, Prince Mine.

Fig. 4 / *L. bunburii* Bell, paratype GSC 7052B; Bell (1938, Pl. 60, Fig. 4); roof of Emery seam. By permission Geological Survey of Canada.
See Fig. 2 and Plate 22, Fig. 1.

Plate 22

Fig. 1 / *Linopteris obliqua* (Bunbury) var. *bunburii* Bell (= *L. bunburii*), holotype GSC 2762; Bell (1938, Pl. 60, Fig. 7). False Bay Lake, below Tracy seam (p. 34). By permission Geological Survey of Canada. This specimen compares with Bell's Fig. 5 (1966, Pl. 22). Refer also to Plate 21, Figs. 2 and 4 in this catalogue.

Fig. 2 / *Neuropteris (Mixoneura) flexuosa* Sternberg (F-5), unspecified coal measure in the Morien series (p. 40); partial pinna.

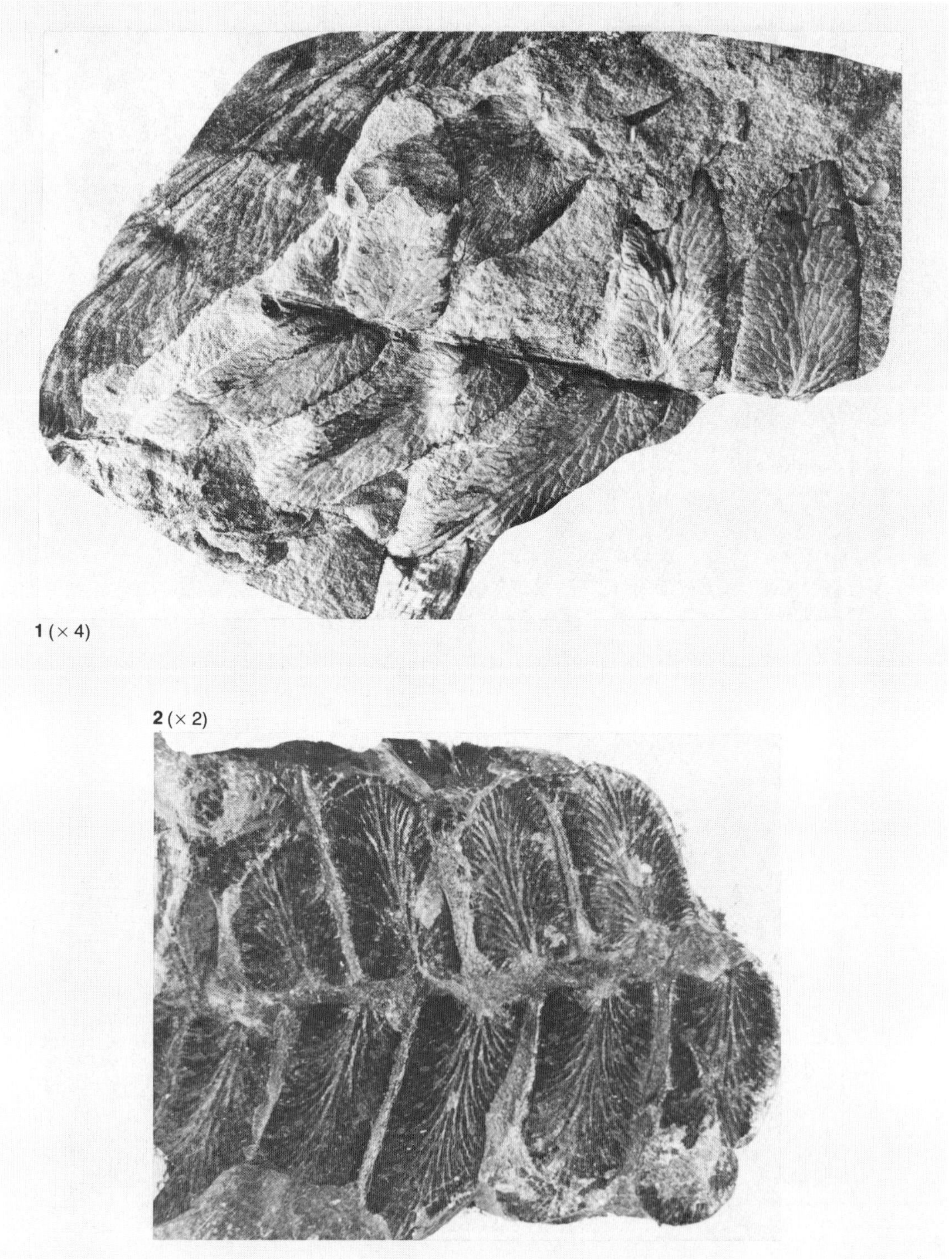

1 (× 4)

2 (× 2)

1 (× 3)

Plate 23

Fig. 1 / *Neuropteris (Mixoneura) flexuosa* Sternberg (F-266). Stubbart seam, Point Aconi, Prince Mine (p. 40); terminal part of a pinna under plane polarized light.

Plate 24

Fig. 1 / *Neuropteris (Mixoneura) flexuosa* Sternberg (F-255). Stubbart seam, Point Aconi, Prince Mine (p. 40). Detail (infrared reflection image) of pinnule in Fig. 3 below.

Fig. 2 / *Neuropteris (Mixoneura) flexuosa* Sternberg (F-281). Same location as in Fig. 1. Infrared reflection image of terminal part of a pinna in Fig. 4 below.

Fig. 3 / *Neuropteris (Mixoneura) flexuosa* Sternberg (F-255). Same location as in Fig. 1. Arrow shows position of pinnule enlarged in Fig. 1 above.

Fig. 4 / *Neuropteris (Mixoneura) flexuosa* Sternberg (F-281). Same location as in Fig. 1; plane polarized light image of parts of a terminal pinna. See Fig. 2 for an enlarged infrared reflection image.

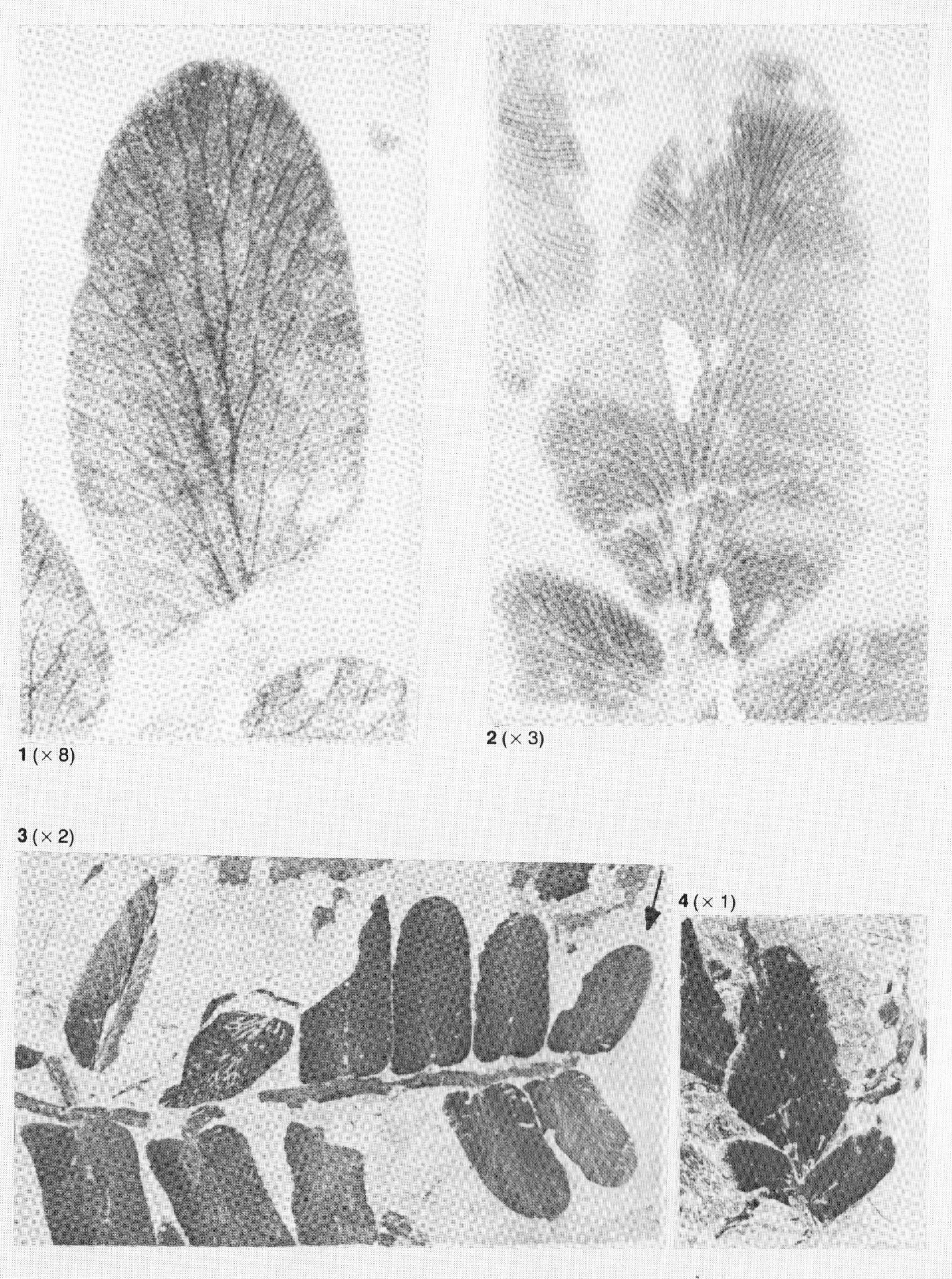

1 (× 2)

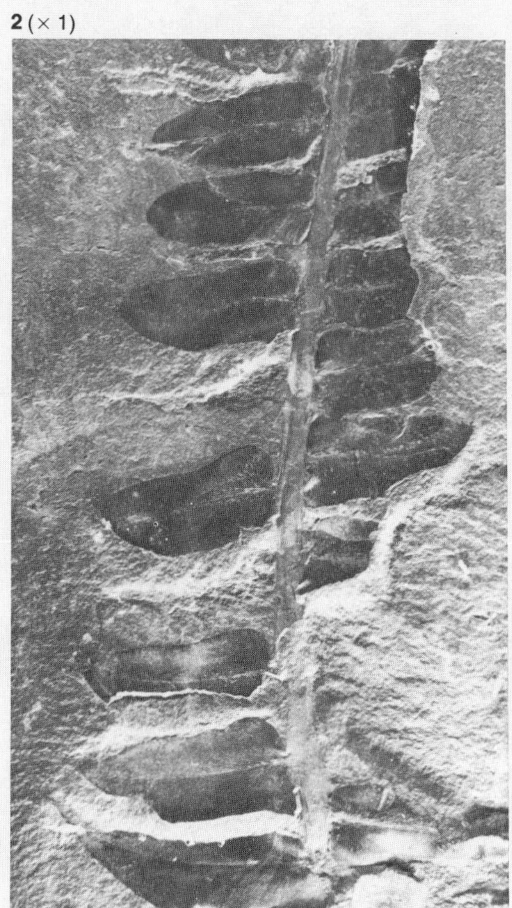
2 (× 1)

3 (× 1.6)

Plate 25

Fig. 1 / *Neuropteris flexuosa* Sternberg forma *magna* (F-375). Harbour seam, Lingan Mine (p. 42); detached pinnule.

Fig. 2 / *Neuropteris flexuosa* Sternberg forma *magna* (F-450). Emery seam, Glace Bay; partial pinna.

Fig. 3 / *Neuropteris flexuosa* Sternberg forma *magna* (F-508). Emery seam, Glace Bay; upper portion of a pinna.

Plate 26

Fig. 1 / *Neuropteris heterophylla* (Brongniart) (F-445). Emery seam, Glace Bay (p. 43); terminal part of a frond. Note the development of pinnae from the top to the bottom of the specimen frond, and the similarity between pinnae and pinnules.

1 (× 2.5)

1 (× 2)

Plate 27

Fig. 1 / *Neuropteris heterophylla* (Brongniart) (F-446). Emery seam, Glace Bay (p. 43); detail of a terminal frond showing two terminal pinnae positioned on the lowest part of the specimen frond.

Plate 28

Fig. 1 / *Neuropteris heterophylla* (Brongniart) (F-68). Mc Aulay seam (p. 43); detached pinnules.

Fig. 2 / *Neuropteris (Mixoneura) obliqua* (Brongniart) (F-474). Phalen seam, Glace Bay (p. 44); apical part of a pinna.

Fig. 3 / *Neuropteris (Mixoneura) ovata* Hoffmann (F-127). Emery seam, Glace Bay (p. 44); apical part of a pinna.

Plate 29

Fig. 1 / *Neuropteris (Mixoneura) ovata* Hoffmann (F-379). Emery seam, Glace Bay (p. 44); a partial frond with a few complete pinnae.

1 (× 1)

Plate 30

Fig. 1 / *Neuropteris (Mixoneura) ovata* Hoffmann (F-129). Emery seam, Glace Bay (p. 44); the mixoneurid condition is illustrated in detail. See *N. (Mixoneura) flexuosa* for explanation (p. 40).

1 (× 4)

1 (× 2)

2 (× 1)

Plate 31

Fig. 1 / *Neuropteris (Mixoneura) ovata* Hoffmann (F-444). Emery seam, Glace Bay (p. 44); infrared reflection image of the upper portion of the specimen pinna.

Fig. 2 / *Neuropteris (Mixoneura) ovata* Hoffmann (F-129). Emery seam, Glace Bay; apical portion of a pinna.

Plate 32

Fig. 1/ *Neuropteris (Mixoneura) ovata* Hoffmann (967G10.66). Cape Breton Island (p. 44); a 15 cm branching stalk with detached fragment of a pinna.

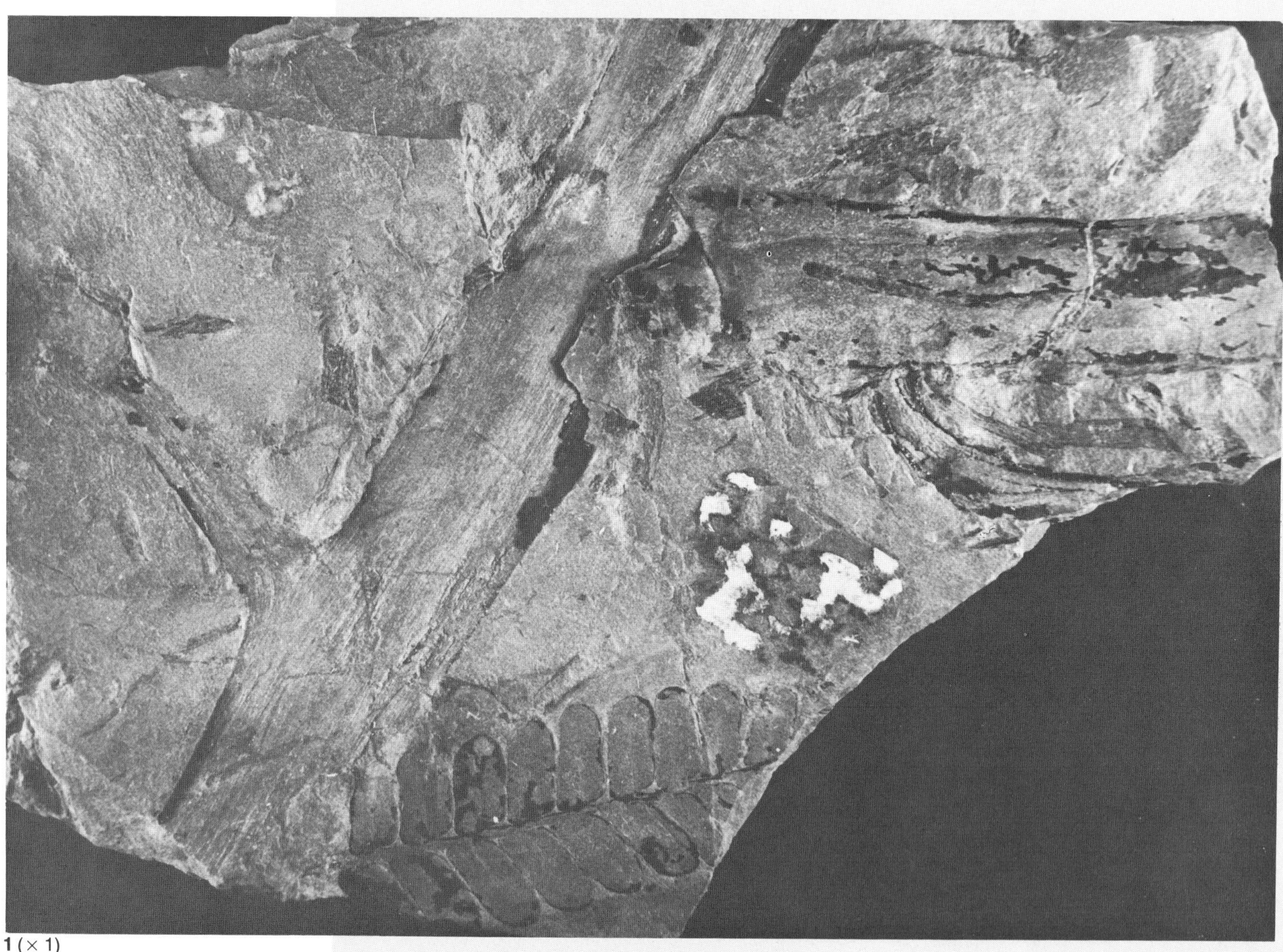

1 (× 1)

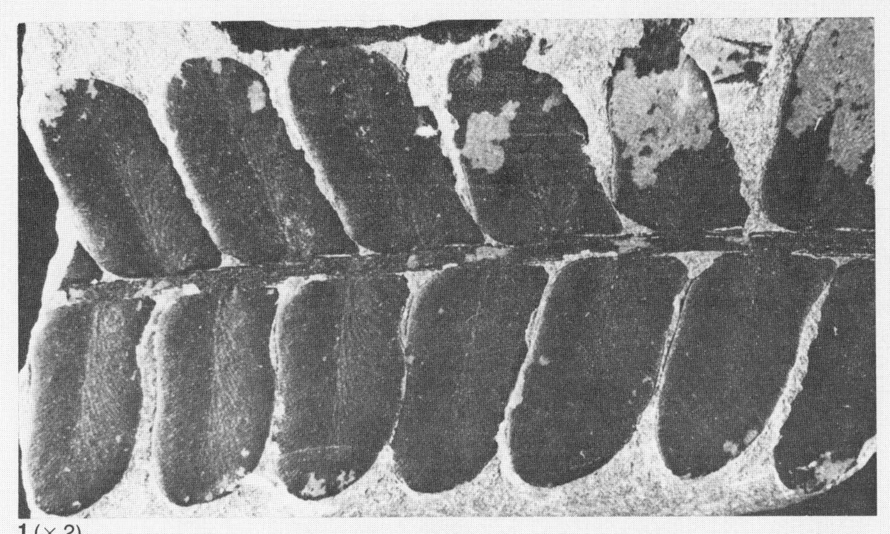

1 (× 2)

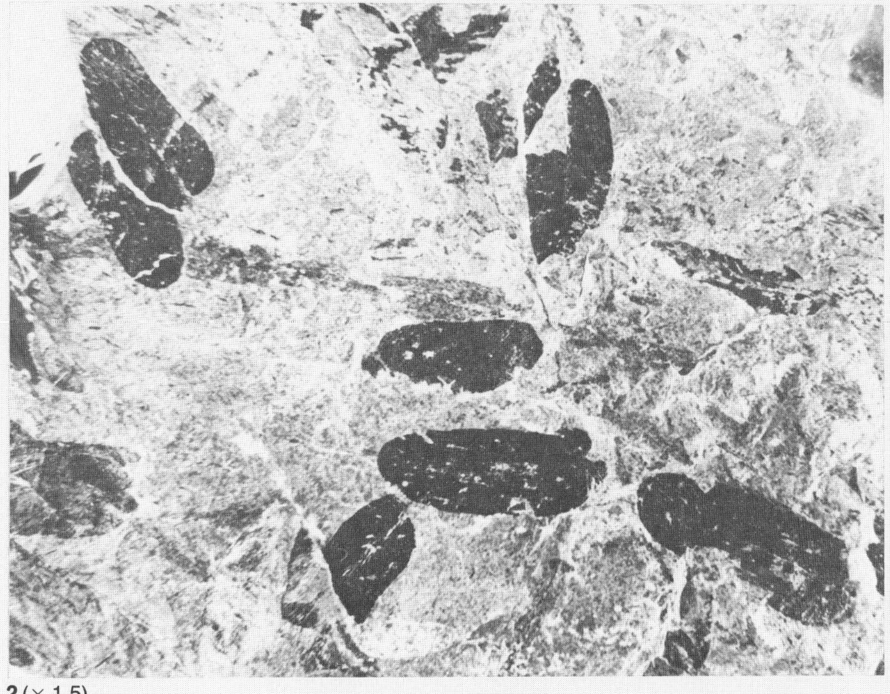

2 (× 1.5)

Plate 33

Fig. 1 / *Neuropteris (Mixoneura) ovata* Hoffmann (F-126). Emery seam, Glace Bay (p. 44); partial pinna.

Fig. 2 / *Neuropteris pseudogigantea* H. Potonié (F-74). Mc Aulay seam, Cape Breton Island (p. 45); detached pinnules.

Plate 34

Fig. 1 / *Neuropteris rarinervis* Bunbury (967G36.1). Cape Breton Island (p. 45); frond with terminal parts missing. For detail refer to Plate 35.

1 (× 1)

1 (× 3)

2 (× 14)

Plate 35

Fig. 1 / *Neuropteris rarinervis* Bunbury (967G36.1). Cape Breton Island (p. 45); detail of two lowermost pinnae. Note mixoneuroid condition; see Fig. 2 for additional detail.

Fig. 2 / *Neuropteris rarinervis* Bunbury (967G36.1). Detail showing characteristic venation pattern of this species. Refer to Fig. 1, (p. 15), and Fig. 6(f)), (g) (p. 41).
(The complete specimen is shown on Plate 34.)

Plate 36

Fig. 1 / *Neuropteris rarinervis* Bunbury (F-363-1). Harbour seam (p. 45); attached pinnules.

Fig. 2 / *Neuropteris rarinervis* Bunbury (F-374). Harbour seam; attached fragments of pinnae.

Fig. 3 / *Neuropteris rarinervis* Bunbury (F-86). Shoemaker seam; entire specimen. Its detail is shown in Plate 37, Fig. 1 (infrared reflection image).

1 (× 6)

3 (× 6)

2 (× 1.5)

1 (× 18)
2 (× 1)
3 (× 1.5)

Plate 37

Fig. 1 / *Neuropteris rarinervis* Bunbury (F-86) detail. Shoemaker seam, Cape Breton Island (p. 45). Infrared reflection image of a part of a pinnule illustrating venation pattern. See Plate 36, Fig. 3 for the entire specimen.

Fig. 2 / *Neuropteris scheuchzeri* Hoffmann (F-13). Unknown coal measure, Cape Breton Island (p. 47); detached pinnule.

Fig. 3 / *Neuropteris scheuchzeri* Hoffmann (F-452). Emery seam, Glace Bay; dichotomizing pinnule. Photographed through a polarizing filter.

Plate 38

Fig. 1 / *Neuropteris scheuchzeri* Hoffmann (F-133). Emery seam, Glace Bay (p. 47).

Fig. 2 / *Neuropteris scheuchzeri* Hoffmann (F-136). Same location as above.

Fig. 3 / *Neuropteris scheuchzeri* Hoffmann (F-316). Harbour seam, Lingan Mine.

All specimens on this plate were collected as detached pinnules.

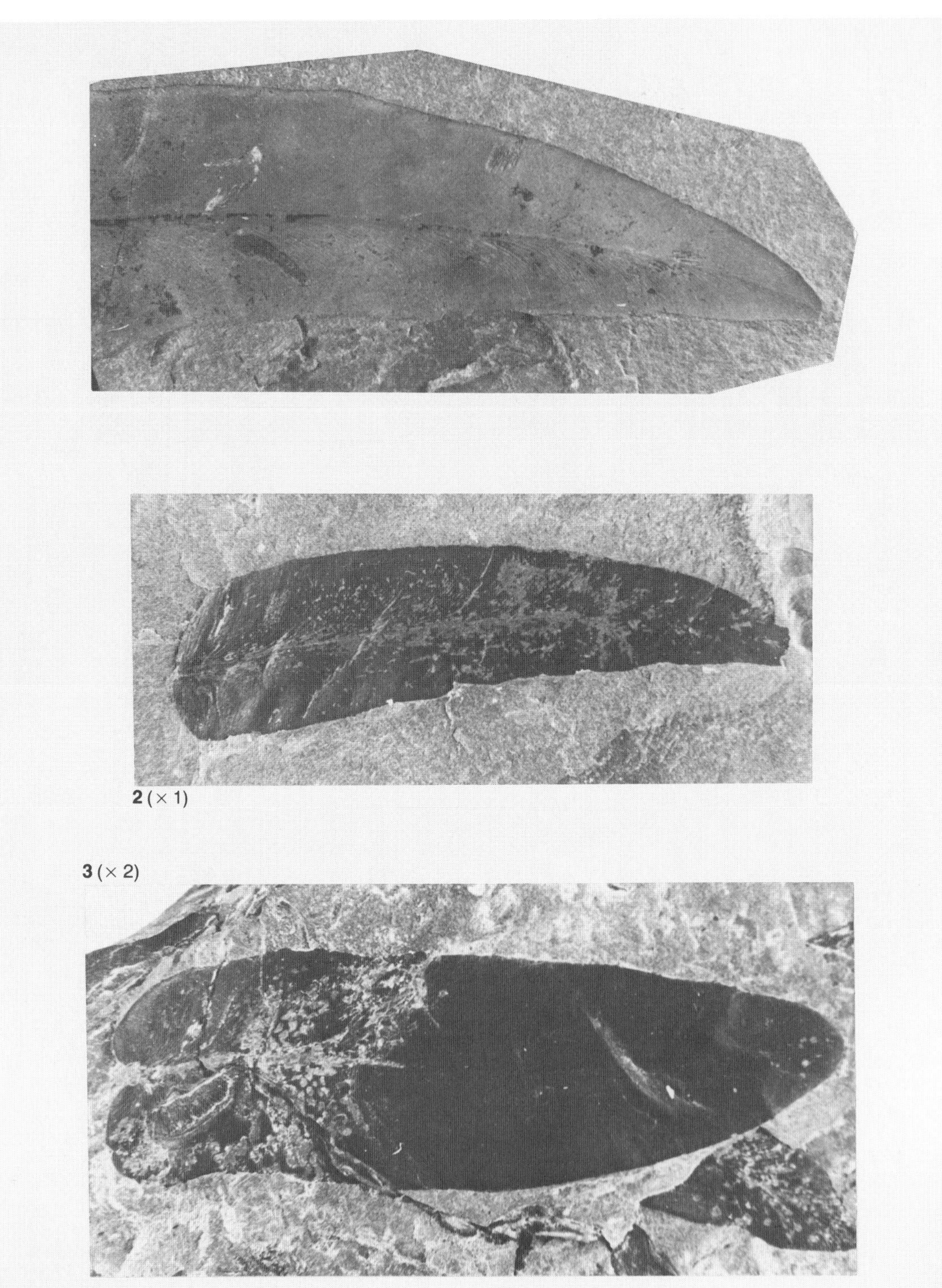

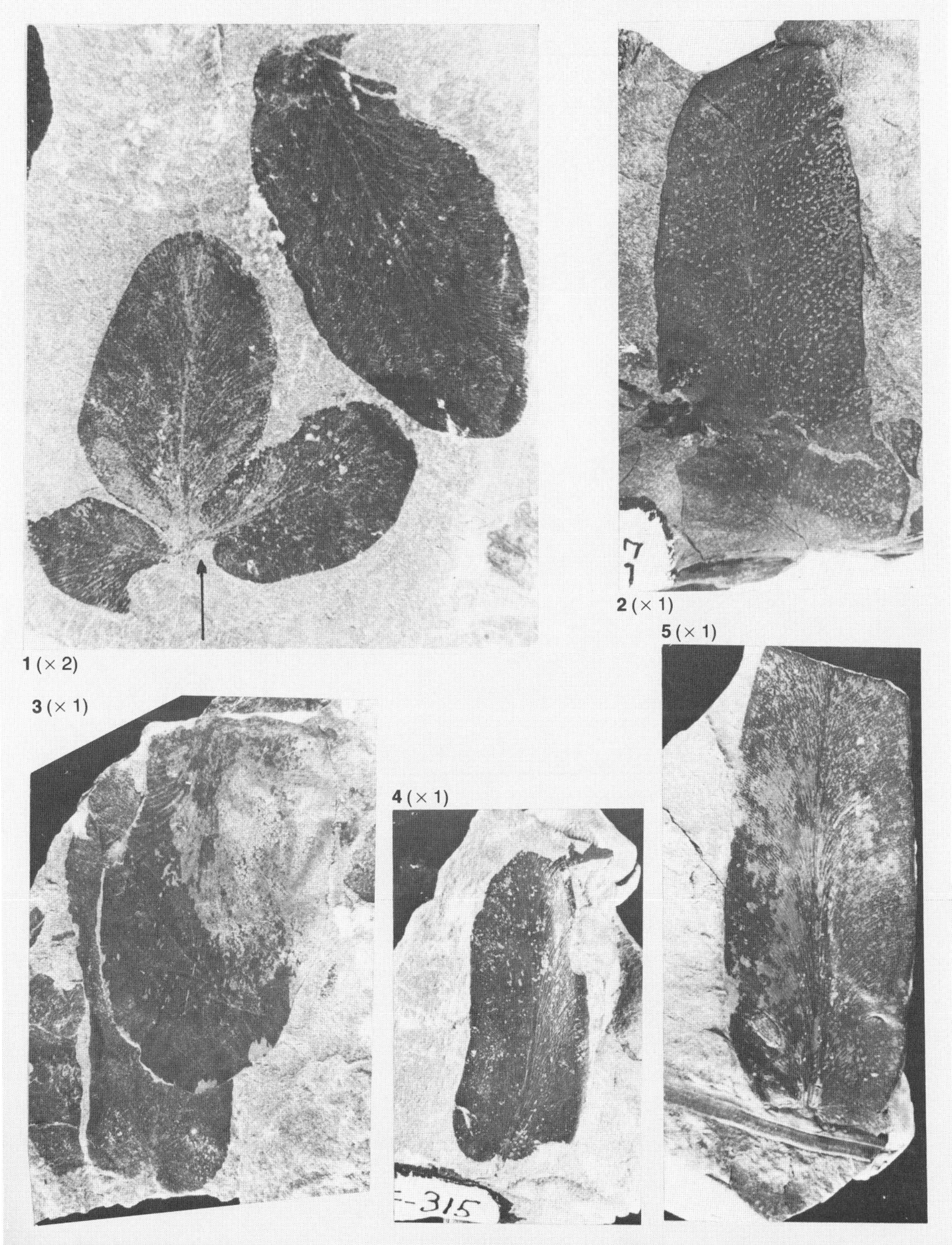

Plate 39

Fig. 1 / *Neuropteris scheuchzeri* Hoffmann (F-449). Emery seam, Glace Bay (p. 47); terminal parts of a simple-pinnate frond. Note the absence of the mixoneurid condition (arrow).

Fig. 2 / *Neuropteris scheuchzeri* Hoffmann (F-317). Harbour seam, Lingan Mine; detached pinnule.

Fig. 3 / *Neuropteris scheuchzeri* Hoffmann (F-318). Harbour seam, Lingan Mine; detached pinnule.

Fig. 4 / *Neuropteris scheuchzeri* Hoffmann (F-315). Harbour seam, Lingan Mine; detached pinnule.

Fig. 5 / *Neuropteris scheuchzeri* Hoffmann (F-387). Emery seam, Glace Bay; attached pinnule.

Plate 40

Fig. 1 / *Neuropteris scheuchzeri* Hoffmann (F-468). Phalen seam, Glace Bay (p. 47).

Fig. 2 / *Neuropteris scheuchzeri* Hoffmann (F-408). Upper (?) Bonar seam, Point Aconi.

Fig. 3 / *Neuropteris scheuchzeri* Hoffmann (F-410). Upper (?) Bonar seam, Point Aconi.

Fig. 4 / *Neuropteris scheuchzeri* Hoffmann (F-564). Emery seam, Glace Bay.

All specimens on this plate were collected as detached pinnules.

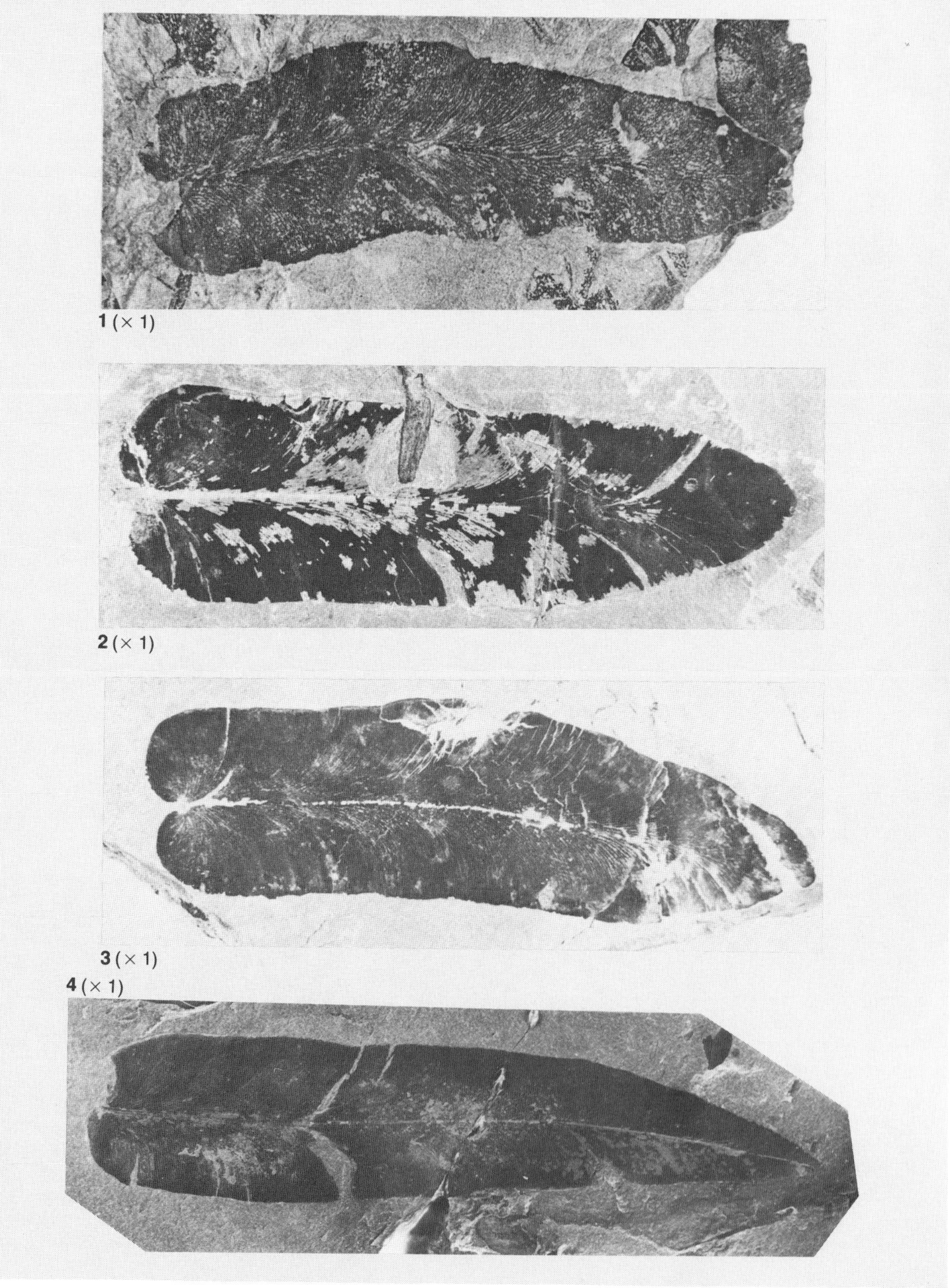

Plate 41

Fig. 1 / *Neuropteris scheuchzeri* Hoffmann (F-501). North Sydney, Harbour seam? (p. 47); the pinnules are assumed to have been attached. For details see following plate.

1 (× 1)

Plate 42

Fig. 1 / *Neuropteris scheuchzeri* Hoffmann (F-501). Detail of what is possibly a seed (center of the pinnule, nut-like structure), and illustration of hair structure especially on the right-hand side pinnule; infrared reflection image. See Plate 41 for the entire specimen.

Fig. 2 / *Neuropteris scheuchzeri* Hoffmann (F-453). Emery seam, Glace Bay (p. 47).

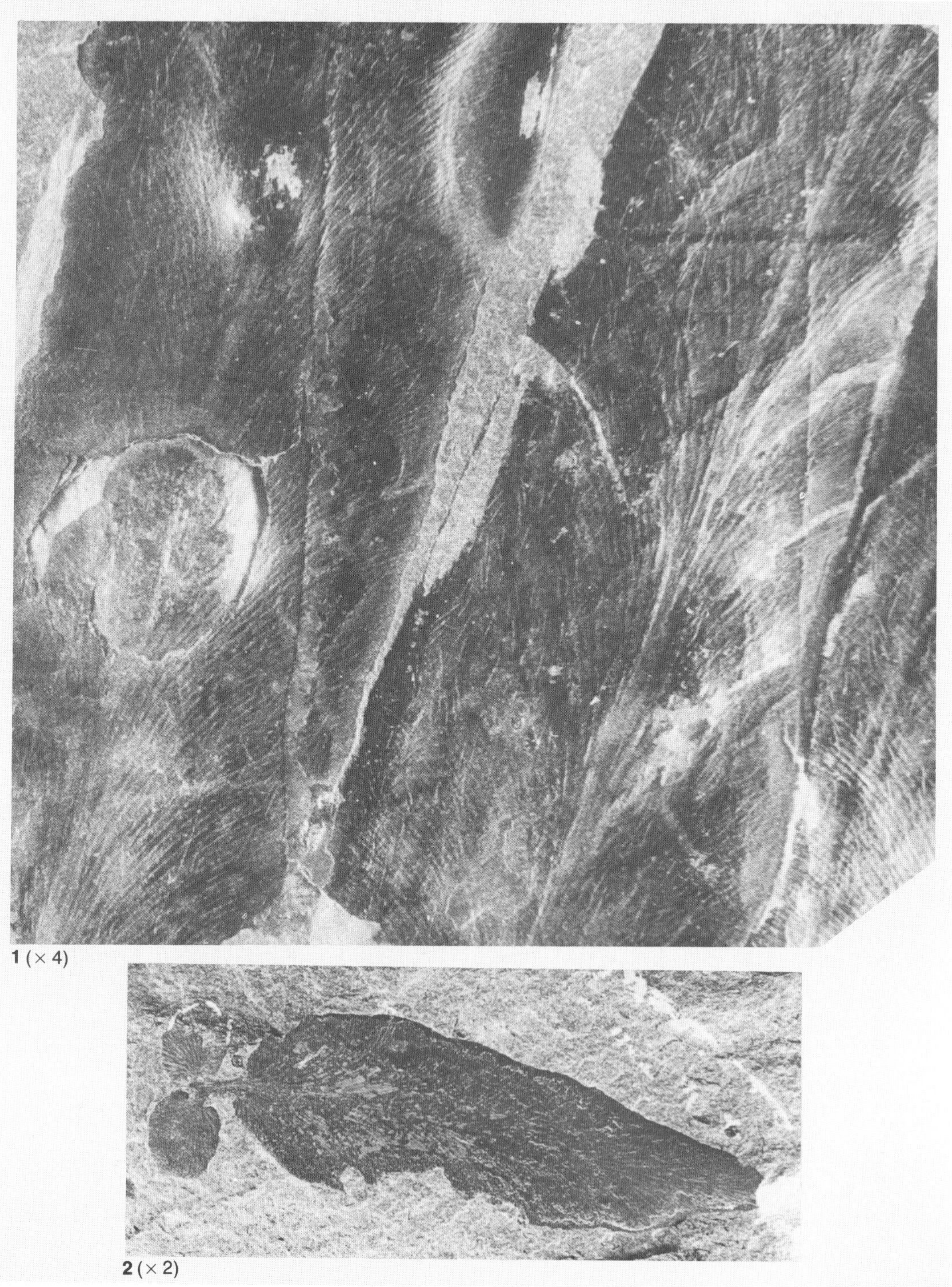

1 (× 4)

2 (× 2)

1 (× 2)

2 (× 1)

Plate 43

Fig. 1 / *Neuropteris scheuchzeri* Hoffmann var. cf. *dawsoni* (F-336). Harbour seam, Lingan Mine (p. 48); detached pinnule.

Fig 2 / *Neuropteris macrophylla* Brongniart (F-637). Harbour seam, Lingan Mine (p. 43); this is an exceptionally good specimen because of the presence of its terminal parts. Note the mixoneuroid condition (in the first pair of pinnules directly below the terminal pinnule).

Plate 44

Fig. 1 / *Neuropteris schlehani* Stur (F-472). Phalen seam, Glace Bay (p. 48); partial pinna.

Fig. 2 / *Neuropteris tenuifolia* (Schlotheim) (F-480). Phalen seam, Glace Bay (p. 49); upper portion of a pinna.

Fig. 3 / Detail of (F-472) in Fig. 1. Third pinnule from bottom right showing venation and mode of attachment of the pinnule including the shape of the base. Note the mixoneurid condition at the lower part of attachment.

1 (× 2)

2 (× 2)

3 (× 10)

Plate 45

Fig. 1 / *Neuropteris* species (F-484). Harbour seam, Lingan Mine (p. 49).

Fig. 2 / *Eremopteris artemisiaefolia* (Sternberg) (F-639). Harbour seam, Lingan Mine (p. 50); infrared reflection image; pinnae are assumed to have been attached.

Fig. 3 / *Eremopteris artemisiaefolia* (Sternberg) (F-622). Stubbart seam, Prince Mine; attached pinnae.

Plate 46

Fig. 1 / *Neuropteris tenuifolia* (Schlotheim) (F-614). Stubbart seam, Prince Mine (p. 49); detail of a frond fragment; attached pinnae (infrared reflection image).

Fig. 2 / *Neuropteris tenuifolia* (Schlotheim) (F-465). Shoemaker seam; entire specimen in the photograph; partial pinna.

Fig. 3 / *Neuropteris tenuifolia* (Schlotheim) (F-185-1). Phalen seam, Glace Bay; partial pinna.

1 (× 2)

2 (× 2)

3 (× 2)

1 (× 2)

2 (× 1.5)

3 (× 1.5)

4 (× 3)

Plate 47

Fig. 1 / *Odontopteris minor* Brongniart (F-503). Phalen seam, Glace Bay (p. 53). Specific identification is difficult if based only on apical parts of a pinna.

Fig. 2 / *Eremopteris artemisiaefolia* (Sternberg) (967G10.4). Morien series, Cape Breton Island (p. 50); partial frond.

Fig. 3 / *Eremopteris artemisiaefolia* (Sternberg) (F-288). Harbour seam, Lingan Mine; upper portion of a pinna.

Fig. 4 / *Eremopteris artemisiaefolia* (Sternberg) (F-105). Tracy seam; terminal parts of a pinna.

Plate 48

Fig. 1 / *Mariopteris latifolia* (Brongniart) (5074). St. Francis Xavier University Collection. (Appendix II, p. 109). Newcastle Mine, N.B.; partial frond. Note *Linopteris muensteri* above "5074".
See Plate 14, Fig. 2 for an enlargement of *L. muensteri*.

1 (× 3)

Plate 49

Fig. 1 / *Mariopteris latifolia* (Brongniart) (F-494). Harbour seam, Lingan Mine (p. 51); attached pinnae.

Fig. 2 / *Mariopteris nervosa* (Brongniart) (F-438). Harbour seam, Lingan Mine (p. 52); upper part of a frond.

Fig. 3 / *Mariopteris nervosa* (Brongniart) (F-469). Phalen seam, Glace Bay; upper part of a frond.

Plate 50

Fig. 1 / *Mariopteris nervosa* (Brongniart) (F-368). Harbour seam, Lingan Mine (p. 52); attached pinnae.

1 (× 2)

1 (× 2)

2 (× 20)

Plate 51

Fig. 1 / *Mariopteris tenuis* Bell (967G10.16). Morien series, Cape Breton Island (p. 52); partial frond.

Fig. 2 / Detail of Fig. 1, showing characteristic venation and shape of the pinnule under infrared reflection.

Plate 52

Fig. 1 / *Odontopteris minor* Brongniart (967G10.25). Mabou Coal Mines, Cape Breton Island (p. 53); entire specimen with details in figures below.

Fig 2. / Detail of (967G10.25) showing upper part of the specimen.

Fig. 3 / Detail of (967G10.25) showing venation, mode of pinnule attachment and shape of pinnule.

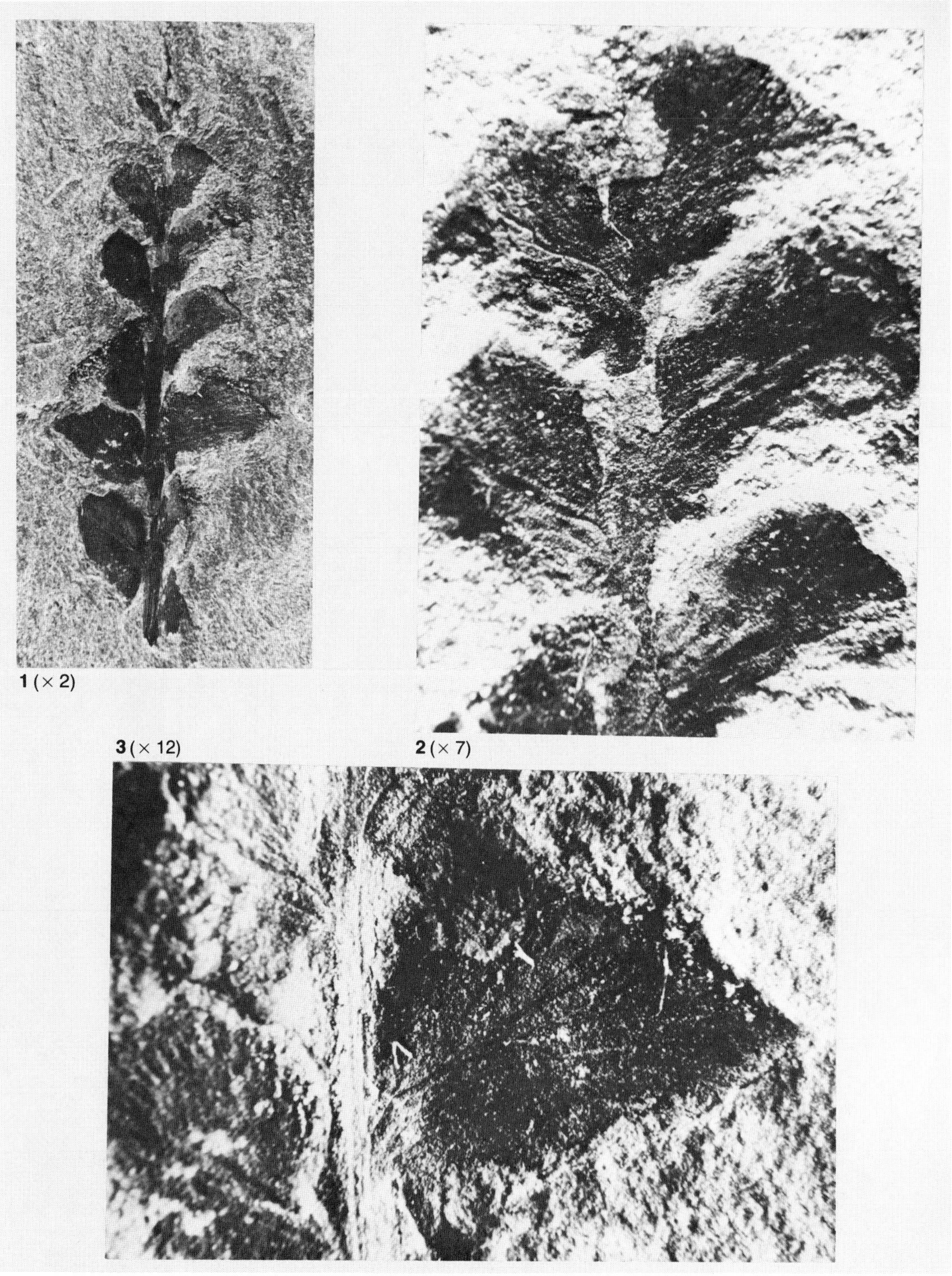

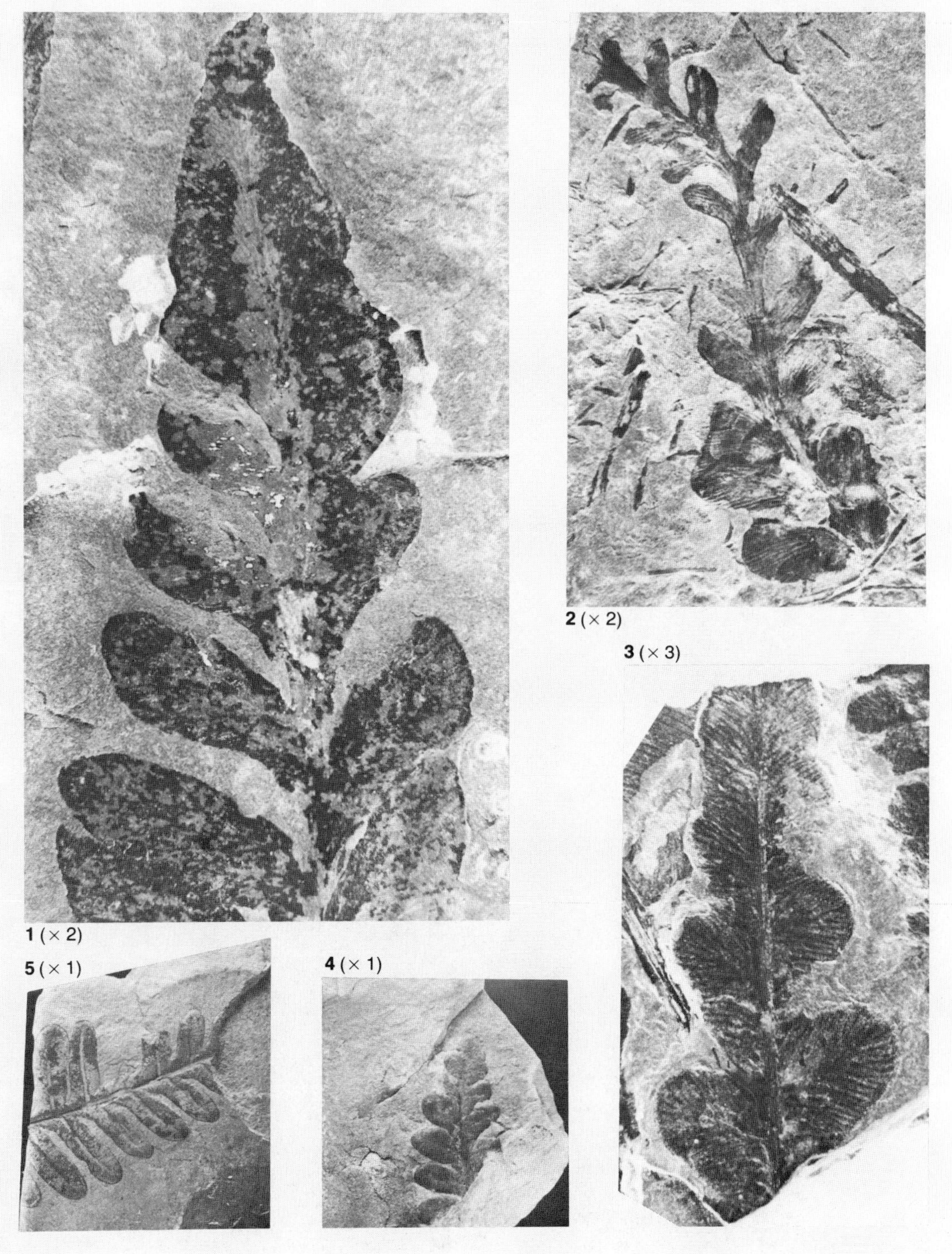

Plate 53

Fig. 1 / *Odontopteris subcuneata* Bunbury (F-580). Emery seam, Glace Bay (p. 53); upper part (?) of a pinna.

Fig. 2 / *Odontopteris subcuneata* Bunbury (F-124). Emery seam, Glace Bay; upper part of a pinna.

Fig. 3 / *Odontopteris subcuneata* Bunbury (F-273). Stubbart seam, Point Aconi; fragment of a pinna.

Fig. 4 / *Callipteridium sullivanti* (Lesquereux) (967G16.7). Harbour seam, Florence Mine (p. 51); apical part of a pinna.

Fig. 5 / *Callipteridium sullivanti* (Lesquereux) (967G16.11). Same location as in Fig. 4; fragment of a pinna.

Plate 54

Fig. 1 / *Odontopteris subcuneata* Bunbury (F-120). Emery seam, Glace Bay (p. 53); detached pinna.

Fig. 2 / Detail of (F-120), marked by arrow in Fig. 1, showing detail of vascular strands entering rachis; infrared reflection image.

Fig. 3 / *Odontopteris subcuneata* Bunbury (967G10.53). Morien series; detached pinna.

In Fig. 1 below arrow is a specimen of *Triletes* cf. *auritus*; see Plate 146, Fig. 3.

1 (× 1)

2 (× 6)

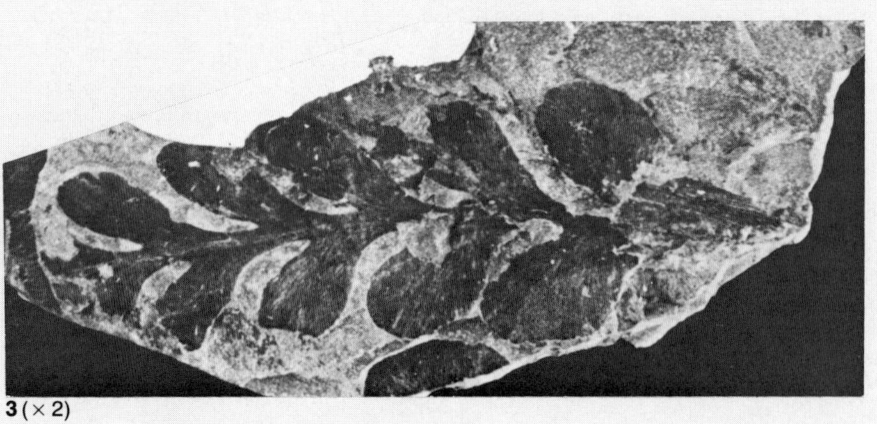

3 (× 2)

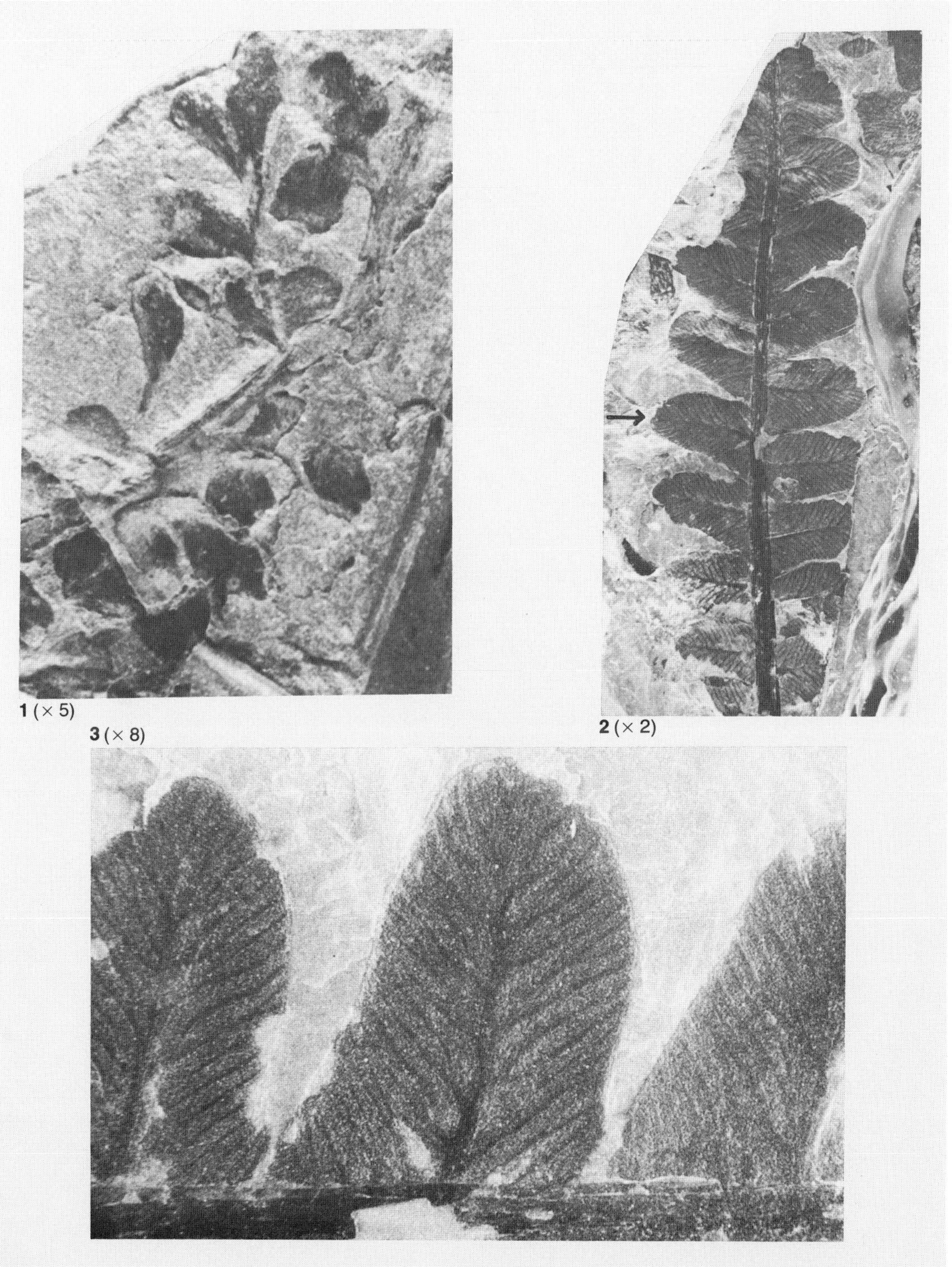

Plate 55

Fig. 1 / *Odontopteris subcuneata* Bunbury (F-87-1). Shoemaker seam (p. 53).

Fig. 2 / *Callipteridium sullivanti* (Lesquereux) (F-275). Stubbart seam, Prince Mine, Point Aconi (p. 51); partial pinna.

Fig. 3 / Detail of (F-275), infrared reflection image, showing venation and decurrent attachment of pinnules to rachis. See arrow in Fig. 2 for position.

Plate 56

Fig. 1 / *Pecopteris* sp. indet. (F-193). Phalen seam, Glace Bay (p. 54); partial pinna.

Fig. 2 / *Sphenopteris neuropteroides* (Boulay) (F-372). Harbour seam, Lingan Mine (p. 54); infrared reflection image of a portion of a frond.

Fig. 3 / *Sphenopteris neuropteroides* (Boulay) (F-157). Emery seam, Glace Bay; portion of a frond.

Fig. 4 / *Sphenopteris neuropteroides* (Boulay) (F-429). Harbour seam, Lingan Mine; portion of a frond.

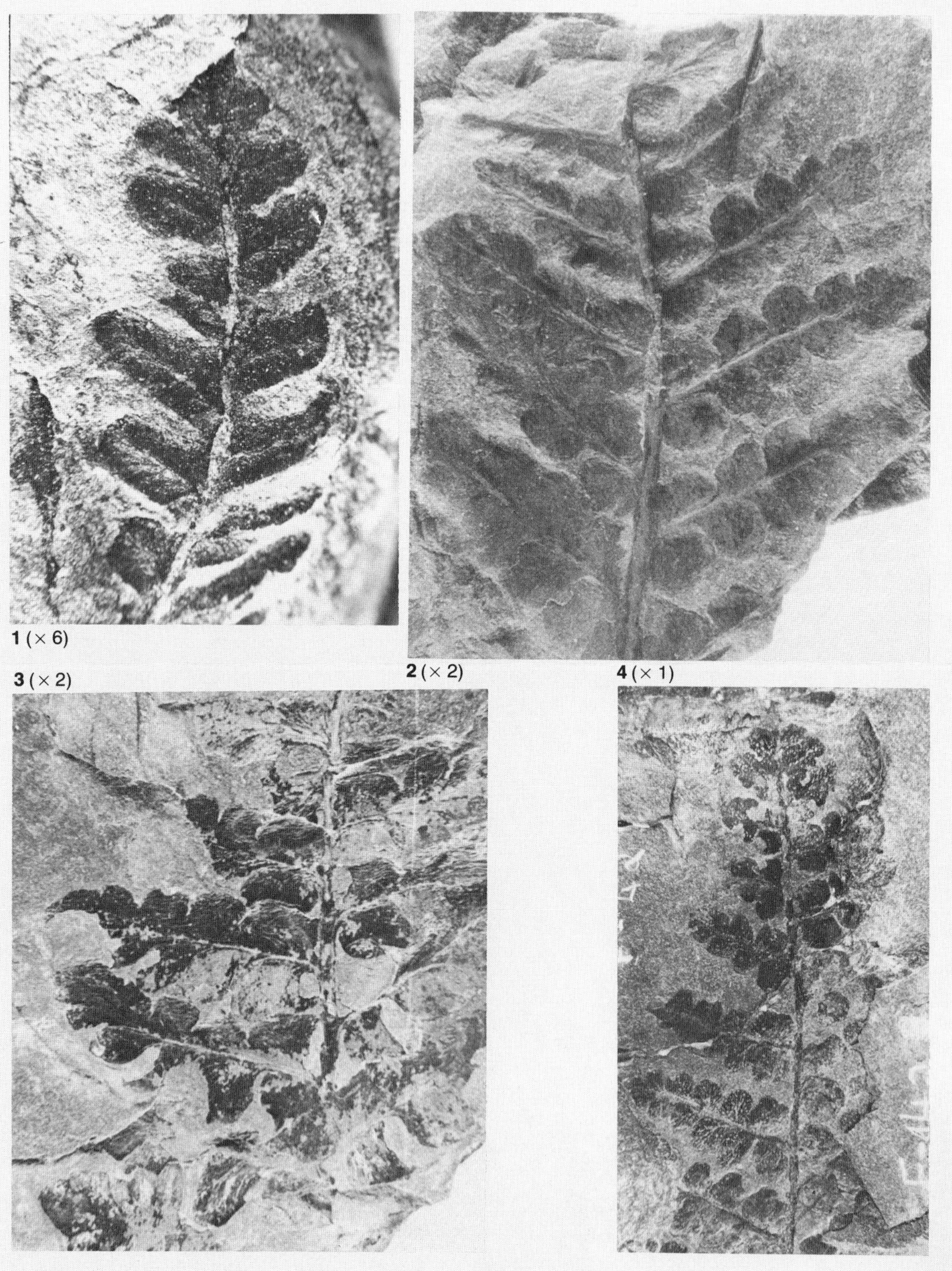

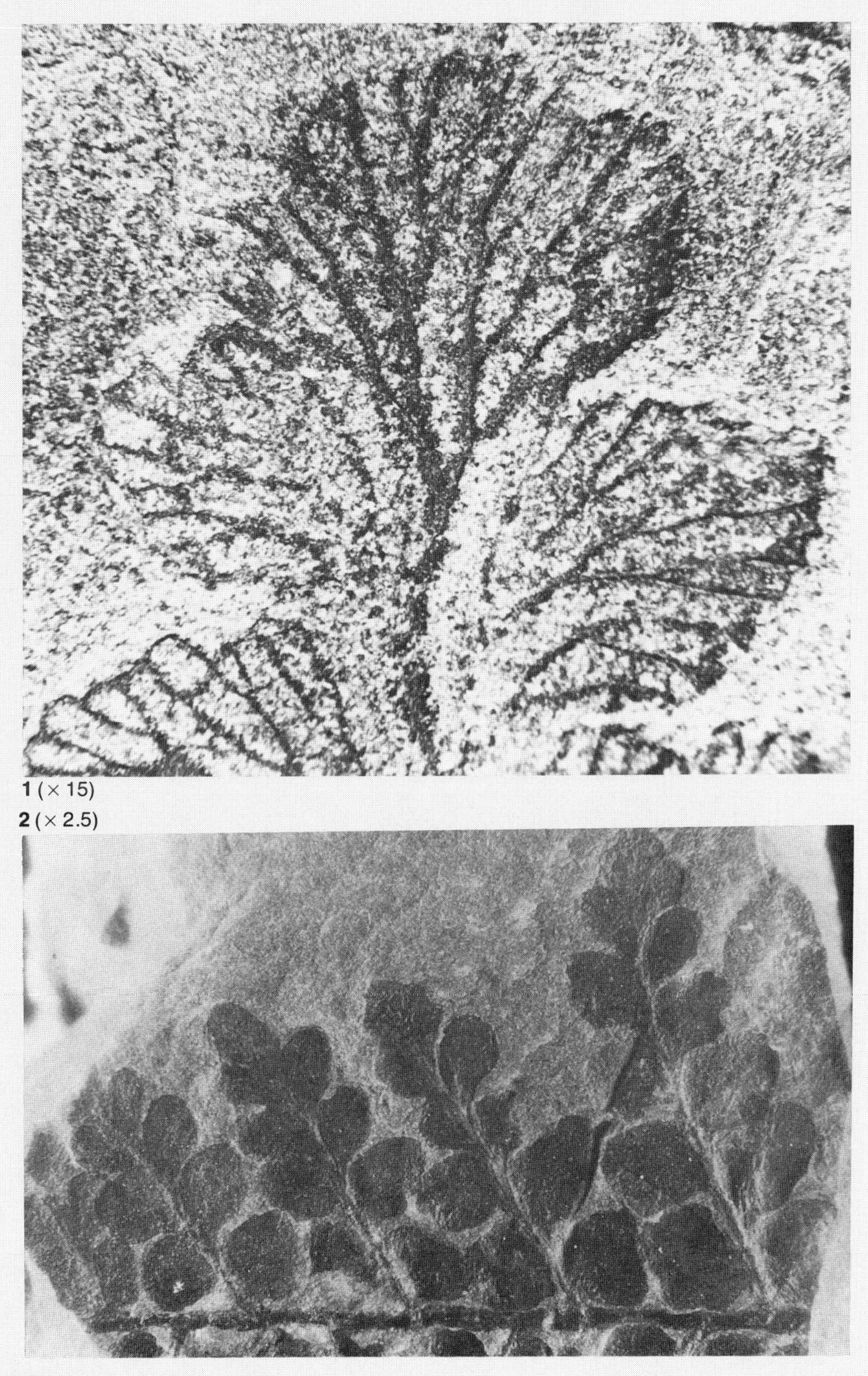

Plate 57

Fig. 1 / *Sphenopteris neuropteroides* (Boulay) (F-22). Harbour seam, #12 Mine, New Waterford (p. 54); apical part of a pinna showing several vascular strands characteristically curved and passing to the pinnule margin.

Fig. 2 / *Sphenopteris neuropteroides* (Boulay) (F-371). Harbour seam, Lingan Mine; infrared reflection image of a partial frond.

Plate 58

Fig. 1 / *Sphenopteris neuropteroides* (Boulay) (F-659). Stubbart seam, Prince Mine (p. 54); partial frond.

1 (× 0.5)

1 (× 1)

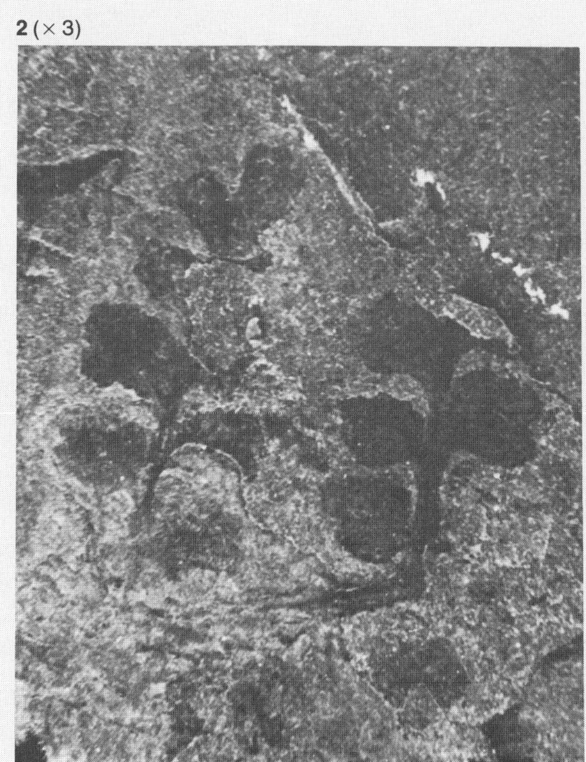

2 (× 3)

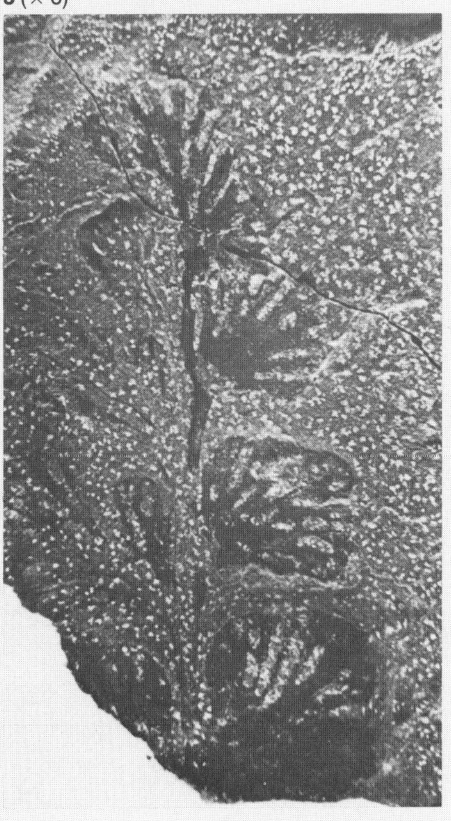

3 (× 6)

Plate 59

Fig. 1 / *Sphenopteris* sp. indet. (F-404). Harbour seam, #26 Mine dump (p. 55); flattened stalk with well developed punctae.

Fig. 2 / *Sphenopteris* cf. *striata* Gothan (F-99). Shoemaker seam (p. 55); attached pinna.

Fig. 3 / *Sphenopteris* cf. *striata* Gothan (F-107). Tracy seam; upper portion of a pinna.

Plate 60

Fig. 1 / *Sphenopteris* cf. *striata* Gothan (F-101). Shoemaker seam (p. 55); apical portion of a pinna.

1 (× 5)

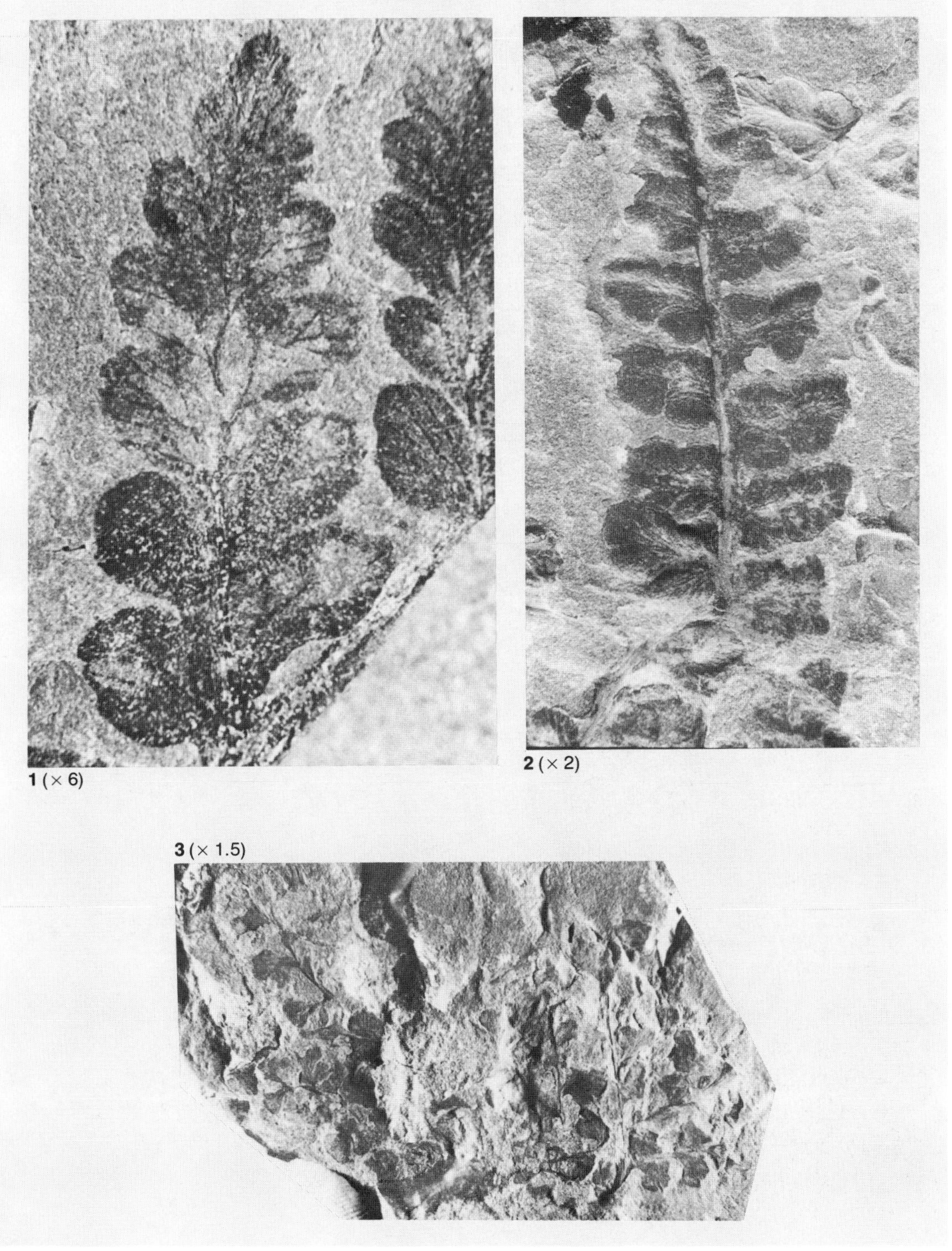

Plate 61

Fig. 1 / *Sphenopteris* cf. *suspecta* D. White (F-412). Upper (?) Bonar seam, Point Aconi (p. 55); detail of a partial frond.

Fig. 2 / *Sphenopteris (Diplotmema) whitii* Bell (F-325). Harbour seam, Lingan Mine (p. 56); infrared reflection image of an apical pinna of a partially preserved frond.

Fig. 3 / *Sphenopteris (Diplotmema) whitii* Bell (F-102). Shoemaker seam; partial frond.

Plate 62

Fig. 1 / *Sphenopteris (Diplotmema) whitii* Bell (F-321). Harbour seam, Lingan Mine (p. 56); a partial frond without terminal parts.

Fig. 2 / *Asterotheca* cf. *abbreviata* (Brongniart) (F-1). Unknown coal seam in the Morien series (p. 56); upper parts of a frond.

1 (× 2)

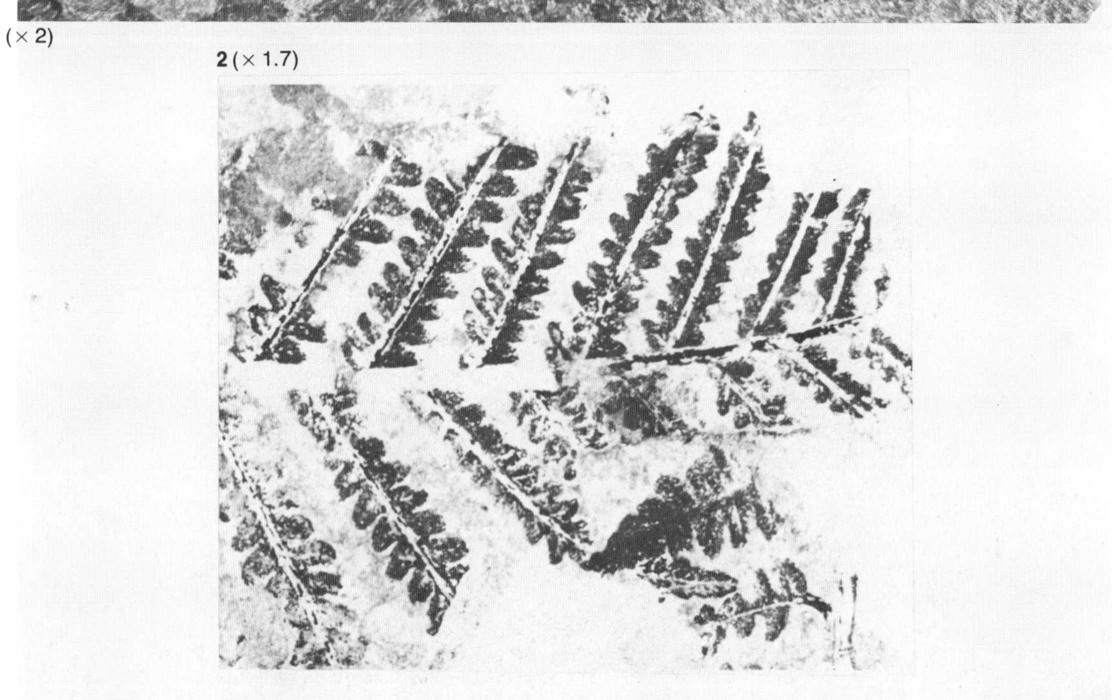

2 (× 1.7)

1 (× 1)

2 (× 3)

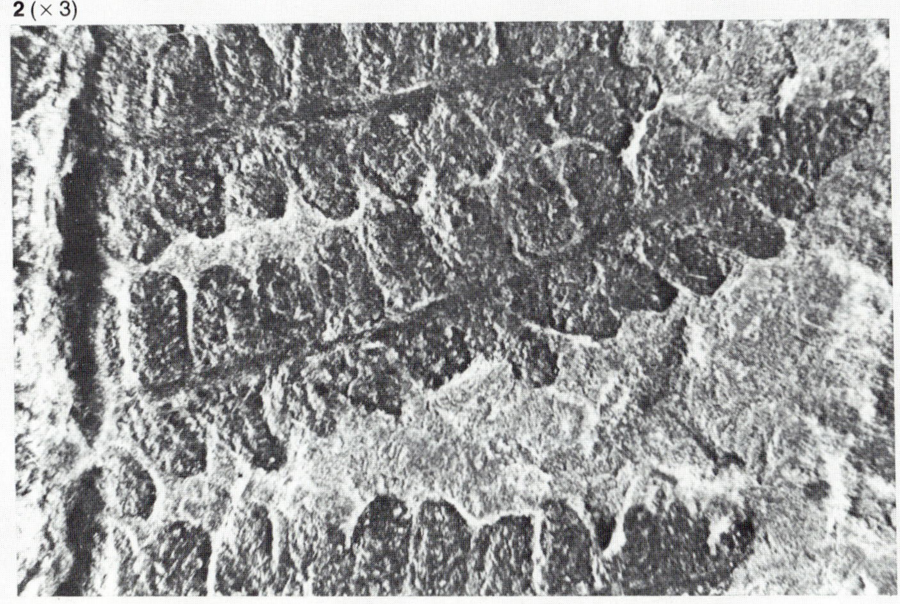

Plate 63

Fig. 1 / *Asterotheca* cf. *abbreviata* (Brongniart) (F-421). Harbour seam, Lingan Mine (p. 56); a fertile frond. Detail shown in Fig. 2.

Fig. 2 / Detail of (F-421) showing impressions of clusters of spore cases (i.e., sori).

Plate 64

Fig. 1 / *Asterotheca daubreei* Zeiller (F-427). Harbour seam, Lingan Mine (p. 57); terminal parts are missing. See following plates for details of this fertile frond. Collected by Dr. C. A. Arnold and K. McCandlish in 1976.

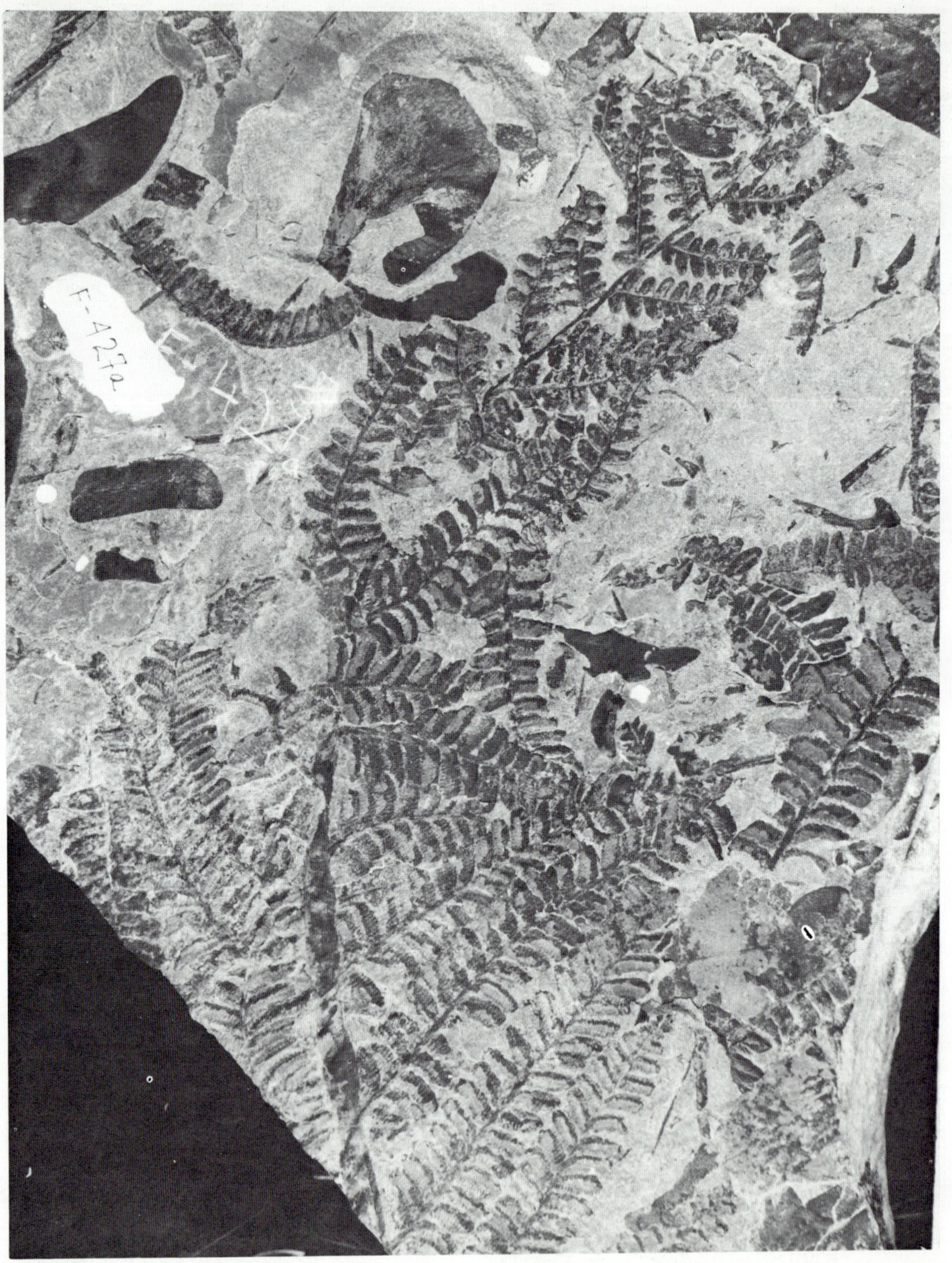

1 (× 0.75)

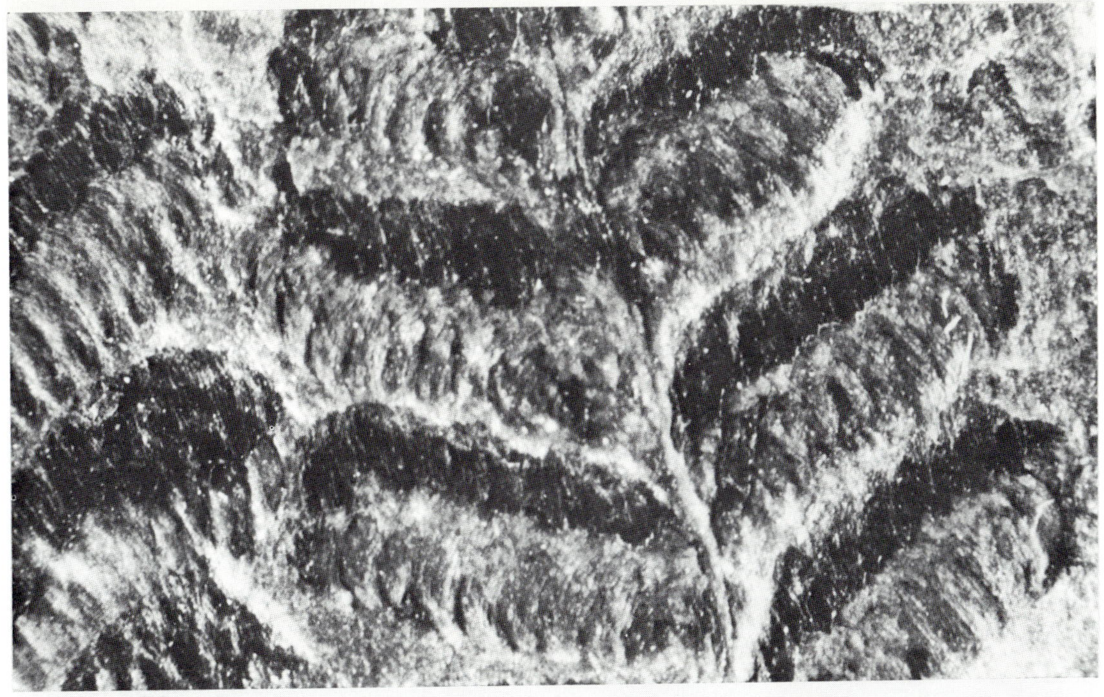

1 (× 5)

2 (× 5)

Plate 65

Fig. 1 / *Asterotheca daubreei.* Detail of (F-427). Infrared reflection image showing pinnules with sori. Note organic carbon remains capping the pinnules, p. 57.

Fig. 2 / *Asterotheca daubreei.* Detail of (F-427) showing sori, some of which are elongate. Note the presence of indentations on the rachis which can be interpreted as punctae. Both details, Figs. 1 and 2, are from the lower right pinnae of the specimen in the previous plate.

Plate 66

Fig. 1 / *Asterotheca daubreei* Zeiller (F-427). Detail of a partial pinna from the upper part, left side, of the frond (p. 57).

Fig. 2 / *Asterotheca herdi* Bell (F-413). Bonar seam, Point Aconi (p. 57); upper part of a pinna.

Fig. 3 / *Asterotheca herdi* Bell (F-287-2). Stubbart seam, Prince Mine, Point Aconi.

1 (× 3)

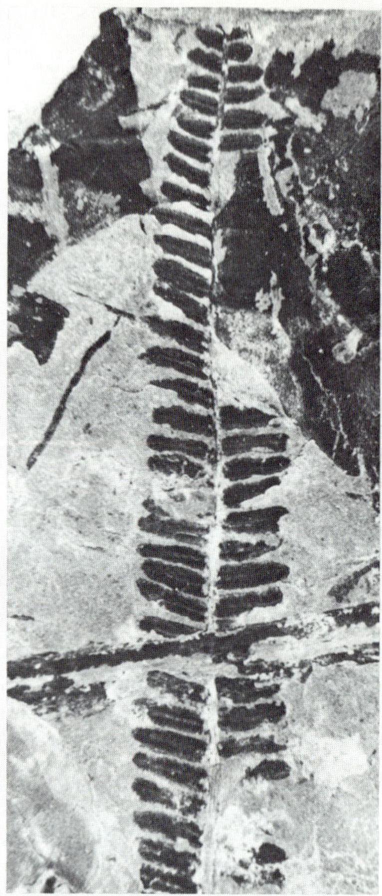

2 (× 2)

3 (× 2)

1 (× 3)

2 (× 1.5)

Plate 67

Fig. 1 / *Asterotheca daubreei* Zeiller (F-432). Harbour seam, Lingan Mine (p. 57); detail of a partial frond.

Fig. 2 / *Asterotheca herdi* Bell (F-287-1). Stubbart seam, Prince Mine, Point Aconi (p. 57); partial pinna.

Plate 68

Fig. 1 / Detail from a partial frond of *Asterotheca herdi* Bell (F-486). Harbour seam, Lingan Mine (p. 57). Shown is a portion of a fertile pinna under infrared reflection. Compare with Fig. 2.

Fig. 2 / Detail from a partial frond of *A. herdi* Bell (F-486). The same portion of a fertile pinna as is shown in Fig. 1. These details are of the lowest portion of the specimen frond (right-hand side).

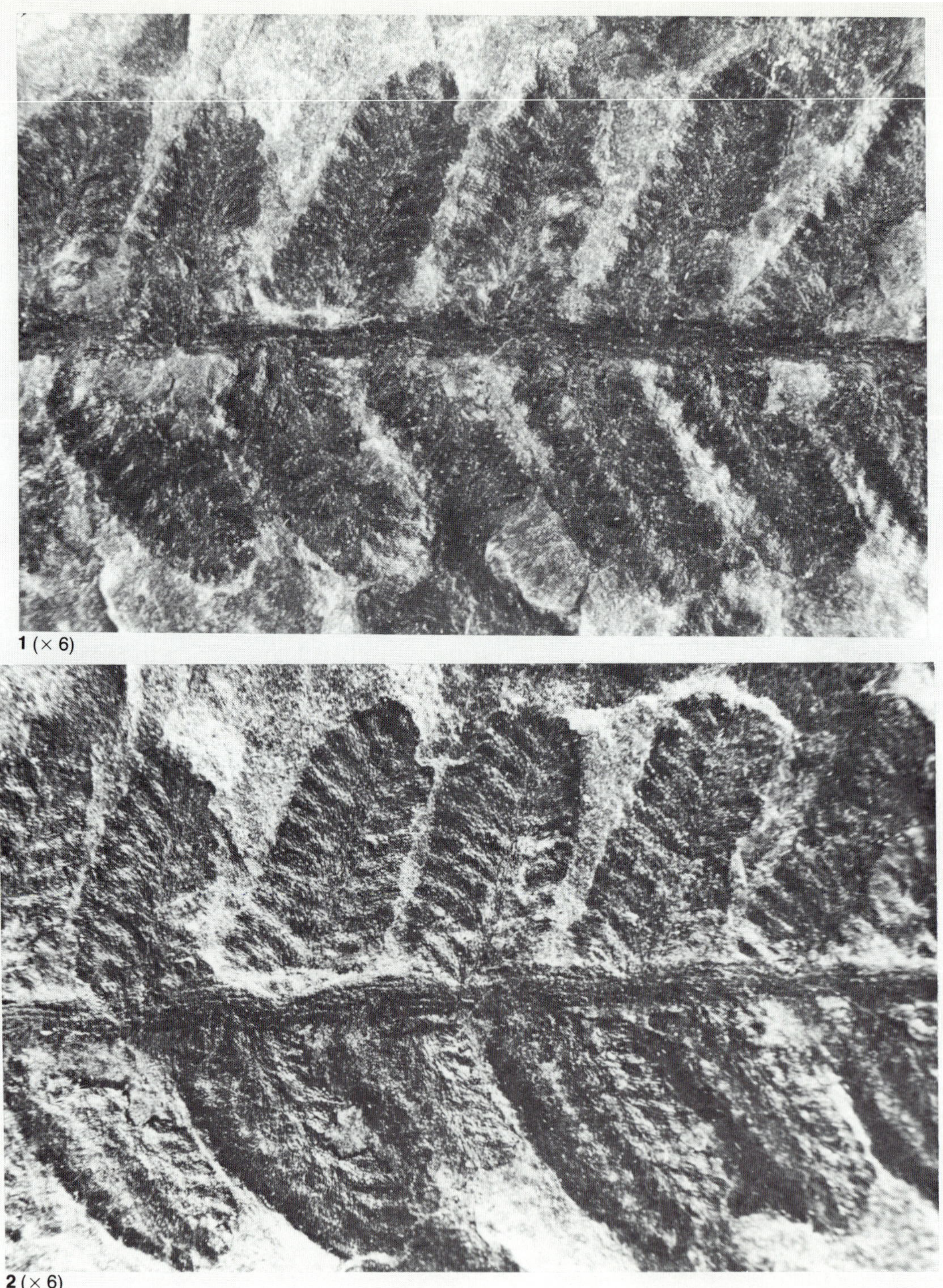

1 (× 6)

2 (× 6)

1 (× 2)

Plate 69

Fig. 1 / *Asterotheca herdi* Bell (F-414). Bonar seam, Point Aconi (p. 57); a partial frond.

Plate 70

Fig. 1 / *Asterotheca oreopteridia* Schlotheim (F-314). Harbour seam, Lingan Mine (p. 58); detail from a partial frond.

1 (× 2)

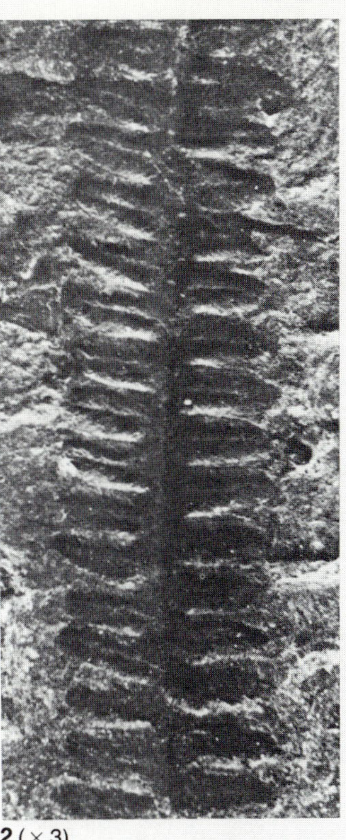

2 (× 3)

3 (× 2)

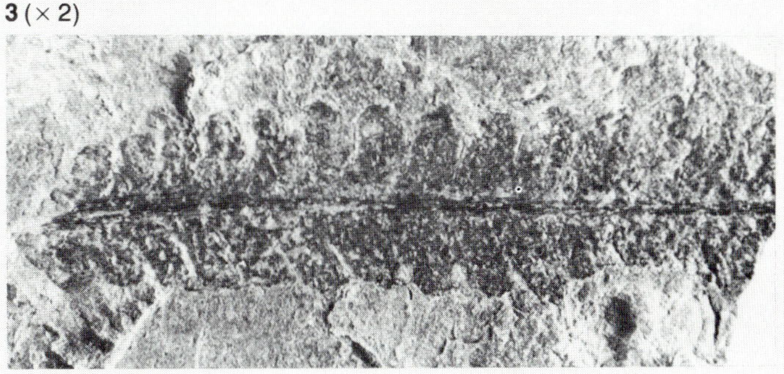

Plate 71

Fig. 1 / *Asterotheca oreopteridia* Schlotheim (F-327). Harbour seam, Lingan Mine (p. 58); partial frond.

Fig. 2 / *Asterotheca oreopteridia* Schlotheim (F-392). Harbour seam, #26 Mine; portion of a pinna.

Fig. 3 / *Asterotheca* sp. indet. (F-415). Bonar seam, Point Aconi (p. 58); portion of a pinna.

Plate 72

Fig. 1 / *Eupecopteris (Asterotheca) cyathea* (Schlotheim) (F-409). Bonar seam, Point Aconi (p. 58); entire specimen (pinna without apical parts). See detail on following plate.

Fig. 2 / *Pecopteris (Asterotheca) acadica* Bell from the Mazon Creek flora in Illinois, Pennsylvanian age; partial frond (p. 59).

Fig. 3 / *Asterotheca* sp. indet. (F-289). Harbour seam, Lingan Mine (p. 58); partial pinna.

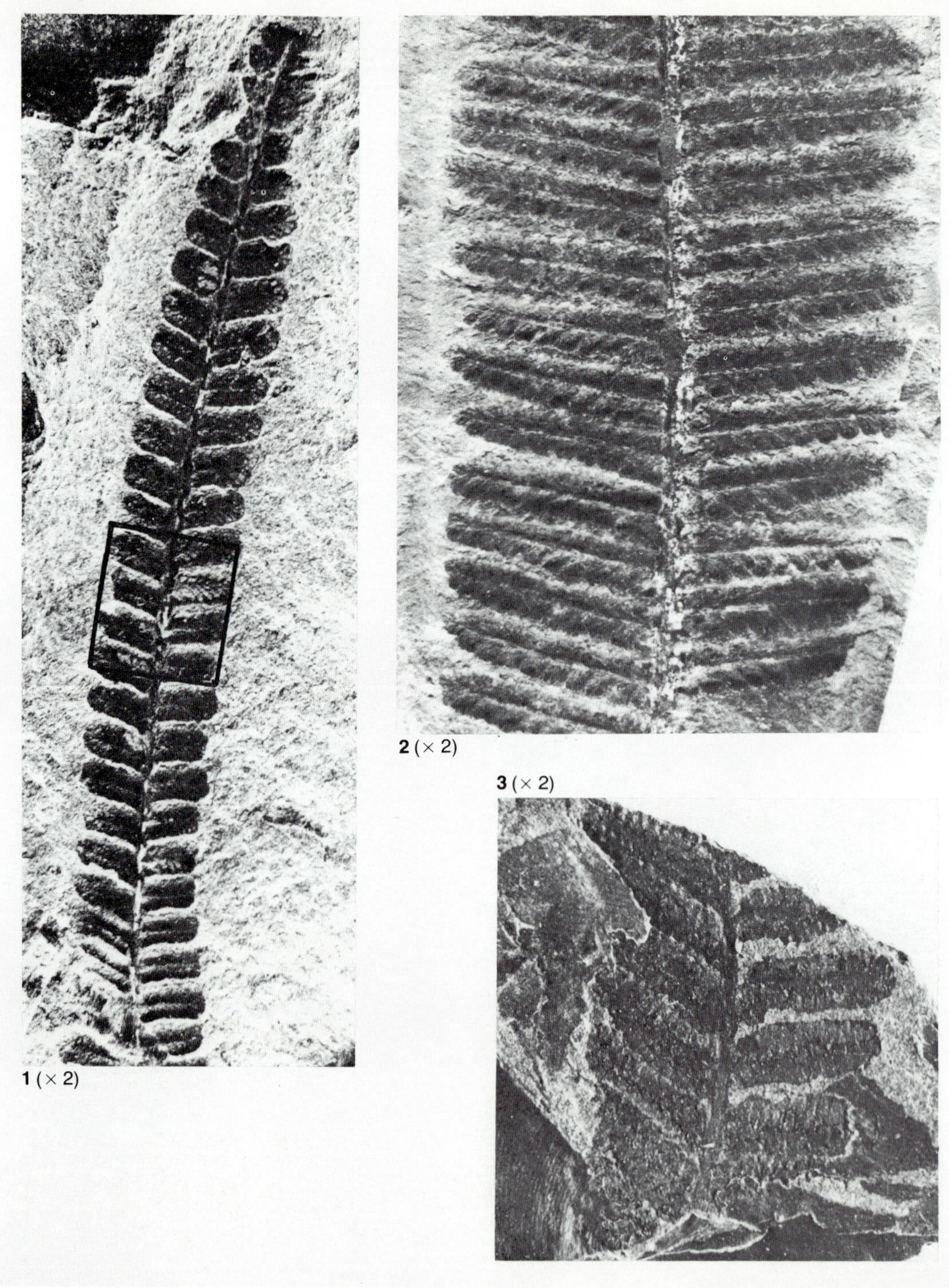

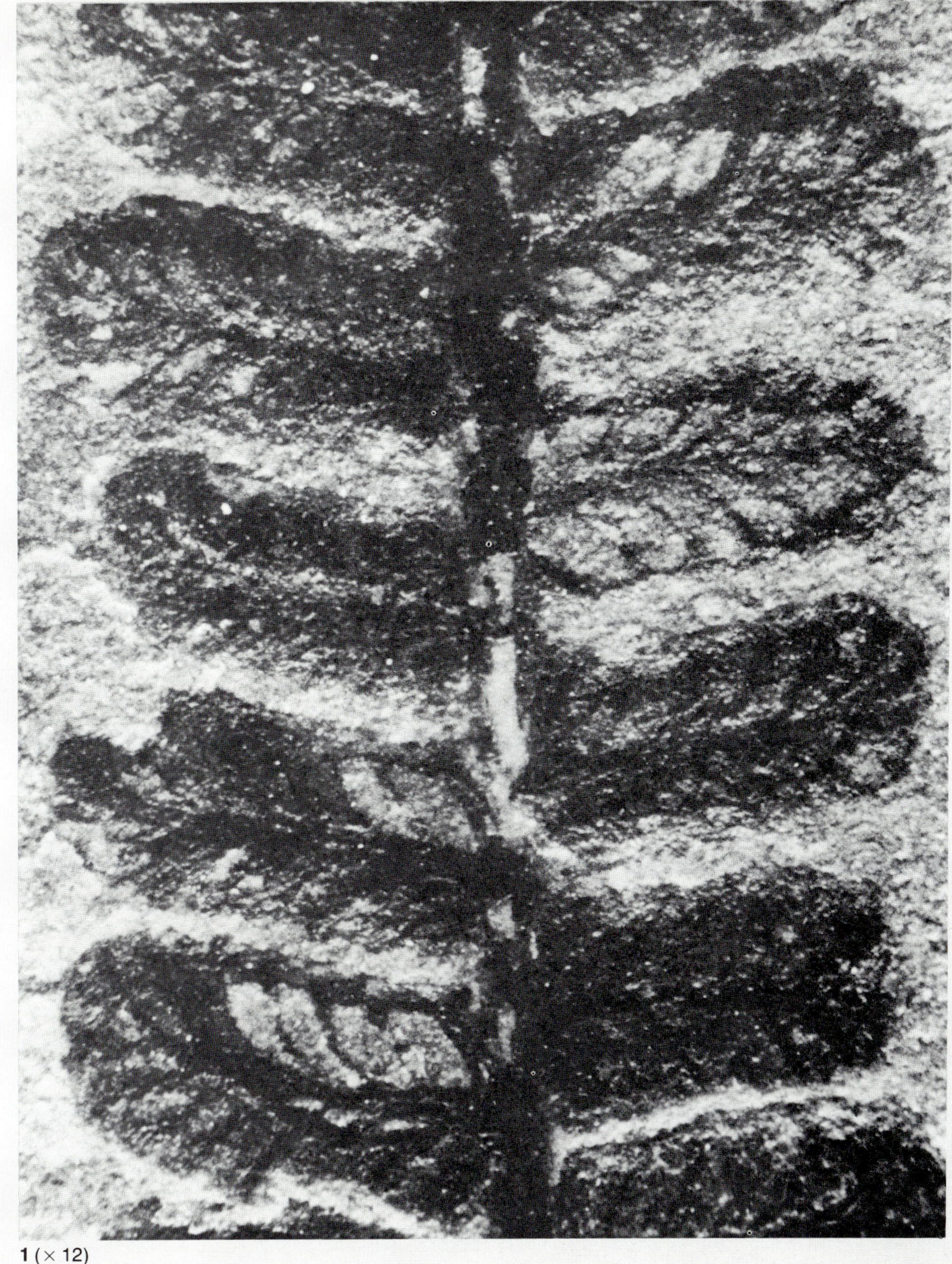

1 (× 12)

Plate 73

Fig. 1 / Detail of *Eupecopteris (Asterotheca) cyathea* (Schlotheim) (F-409), showing arrangement of sori on the pinnules under infrared reflection (p. 58). Refer to the previous plate for a photograph of the entire specimen.

Plate 74

Fig. 1 / *Pecopteris (Asterotheca) acadica* Bell (F-566). Emery seam, Glace Bay (p. 59); terminal parts of a frond.

Fig. 2 / *Pecopteris (Asterotheca) acadica* Bell (F-457). Emery seam, Glace Bay; detail of pinnule morphology from a frond.

1 (× 1)

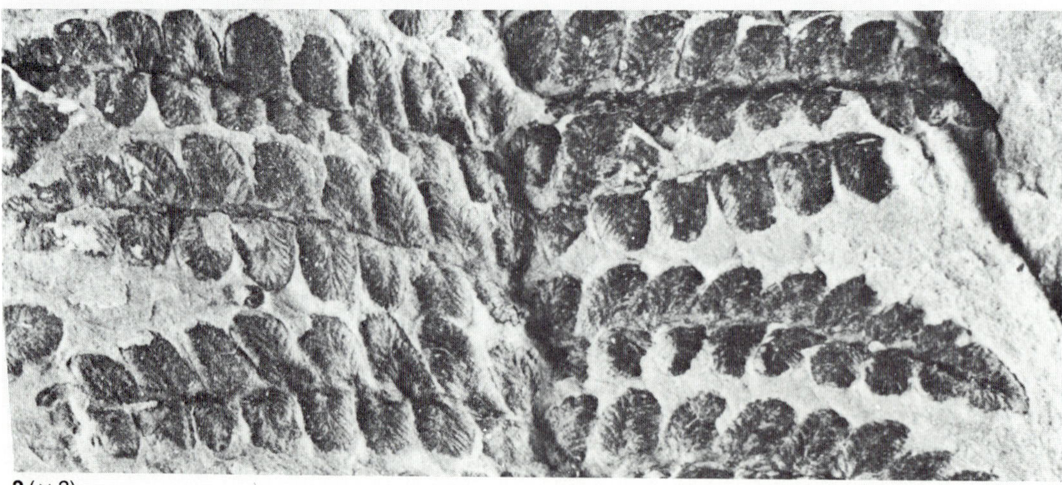

2 (× 3)

1 (× 1)

Plate 75

Fig. 1 / *Pecopteris (Asterotheca) acadica* Bell (F-162). Emery seam, Glace Bay (p. 59); upper parts of a frond.

Plate 76

Fig. 1 / *Pecopteris (Asterotheca) acadica* Bell (F-284). Stubbart seam, Point Aconi (p. 59); upper parts of a frond.

Fig. 2 / *Pecopteris (Asterotheca) acadica* Bell (F-248). Stubbart seam, Prince Mine, Point Aconi; infrared reflection image illustrating venation pattern in pinnules. Note the thick carbon layer.

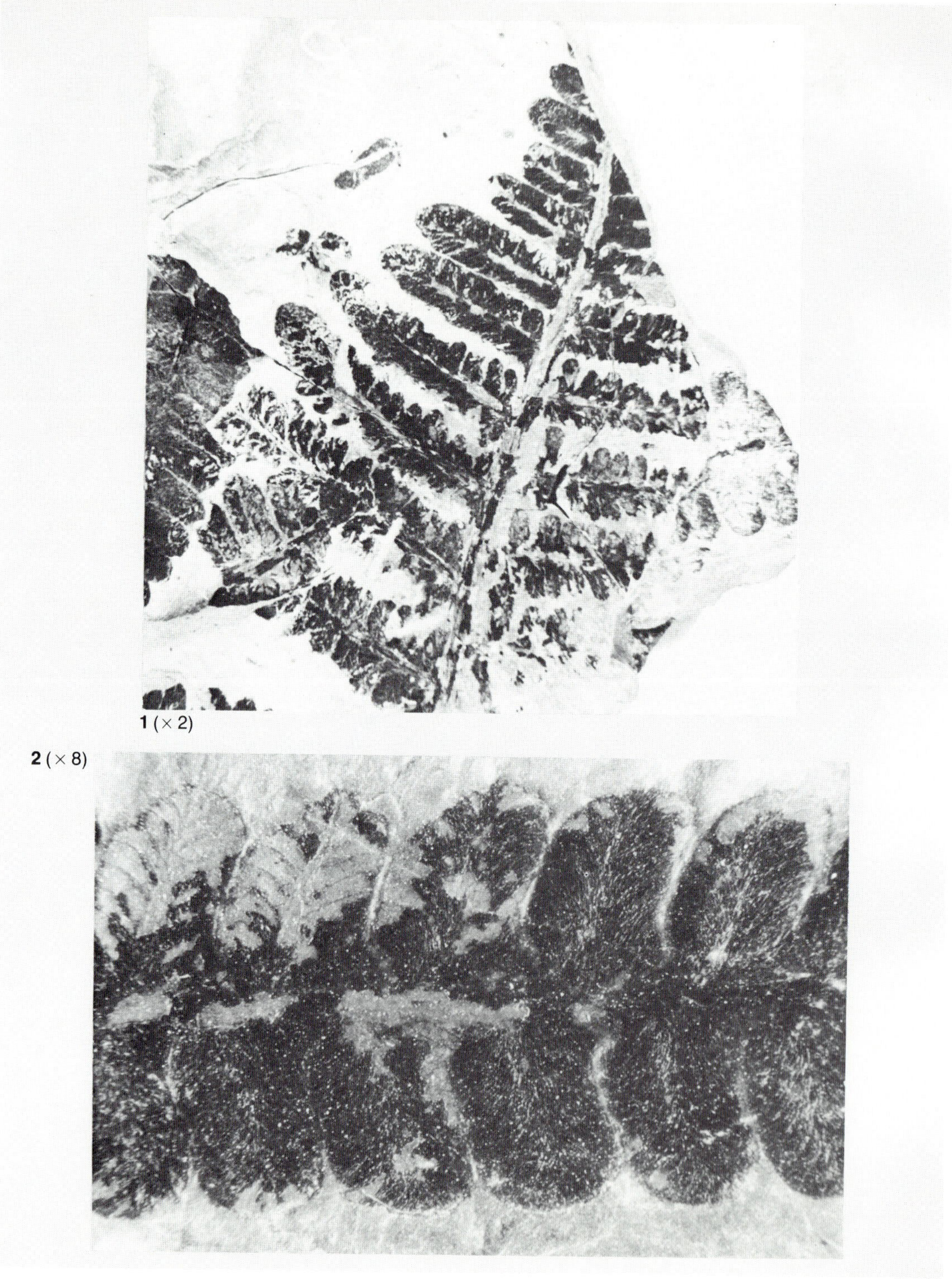

1 (× 1)

2 (× 1.5)

Plate 77

Fig. 1 / *Pecopteris (Asterotheca) acadica* Bell (F-167). Emery seam, Glace Bay (p. 59); upper parts of a frond.

Fig. 2 / *Pecopteris (Asterotheca) acadica* Bell (F-397). Harbour seam, #26 Mine dump; upper portion of a pinna.

Plate 78

Fig. 1 / *Pecopteris (Asterotheca) acadica* Bell (F-169). Emery seam, Glace Bay (p. 59); terminal pinnae on a frond.

Fig. 2 / *Pecopteris (Asterotheca) hemitelioides* Brongniart (F-203). Morien series (p. 60); a pinna with terminal parts.

Fig. 3 / *Pecopteris (Asterotheca) hemitelioides* Brongniart (F-285). Stubbart seam, Prince Mine, Point Aconi; fragments of pinnae.

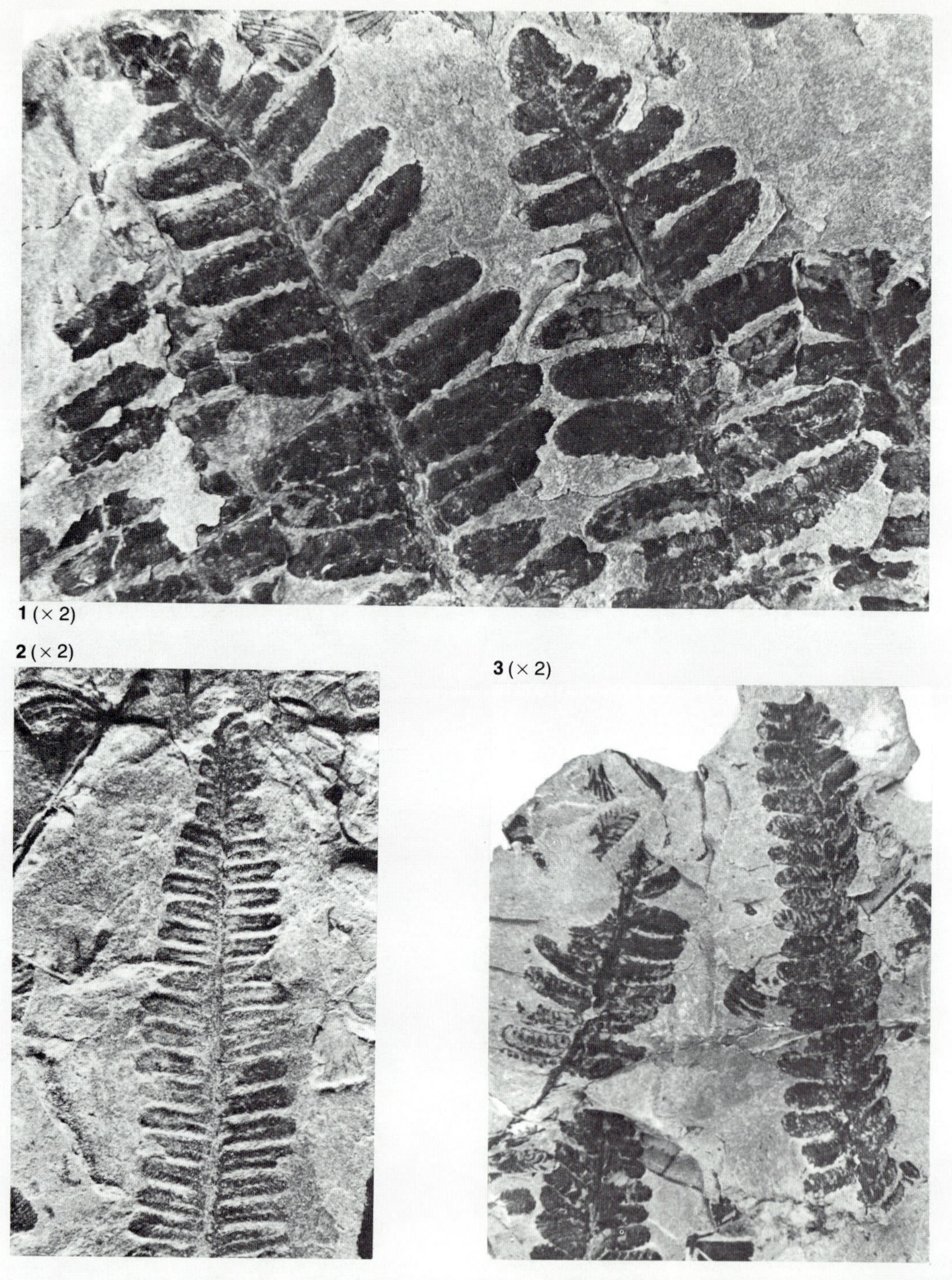

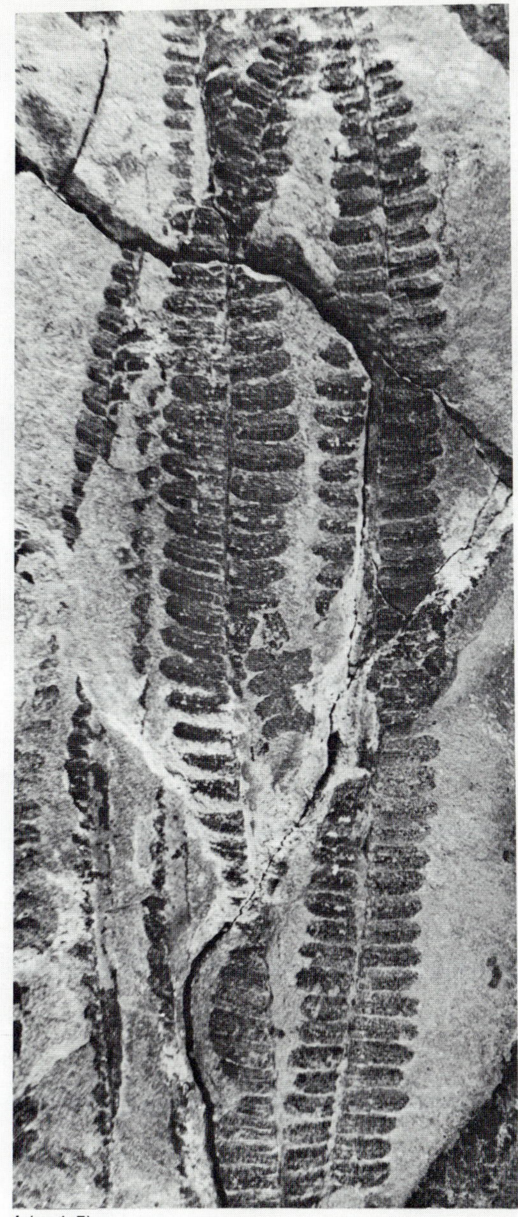

1 (× 1.5)

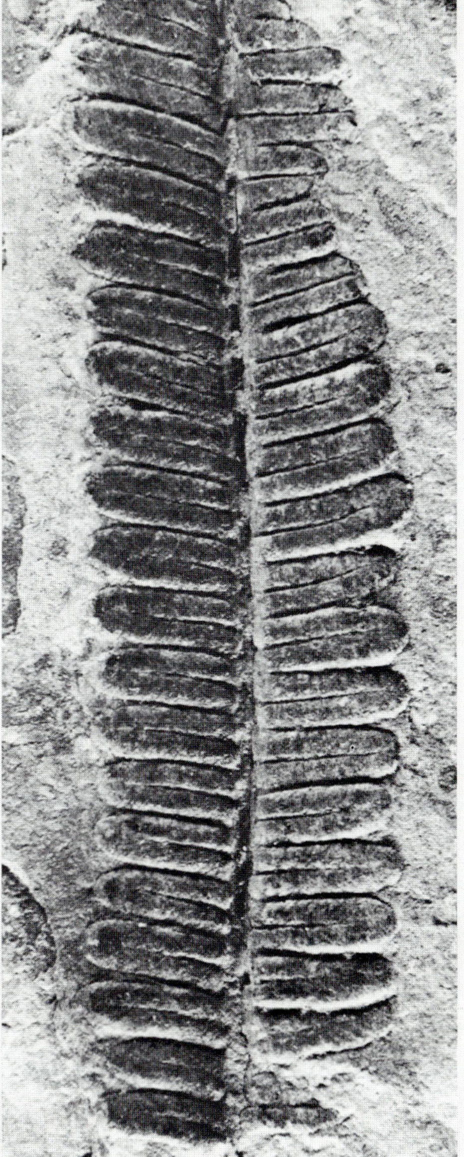

2 (× 4)

Plate 79

Fig. 1 / *Pecopteris (Asterotheca) hemitelioides* Brongniart (F-407). Bonar seam, Point Aconi (p. 60); several partial pinnae.

Fig. 2 / *Pecopteris (Asterotheca) hemitelioides* Brongniart (F-383). Emery seam, Glace Bay; part of a pinna.

Plate 80

Fig. 1 / *Pecopteris (Senftenbergia) pennaeformis* Brongniart (F-589). Emery seam, Glace Bay (p. 60); upper part of a frond with terminal parts missing.

Fig. 2 / *Pecopteris (Senftenbergia) pennaeformis* Brongniart (F-495). Harbour seam, Lingan Mine; detail of feather-like form of pinnae which gives the species its name. Note the sturdy rachis in each pinna.

1 (× 1)

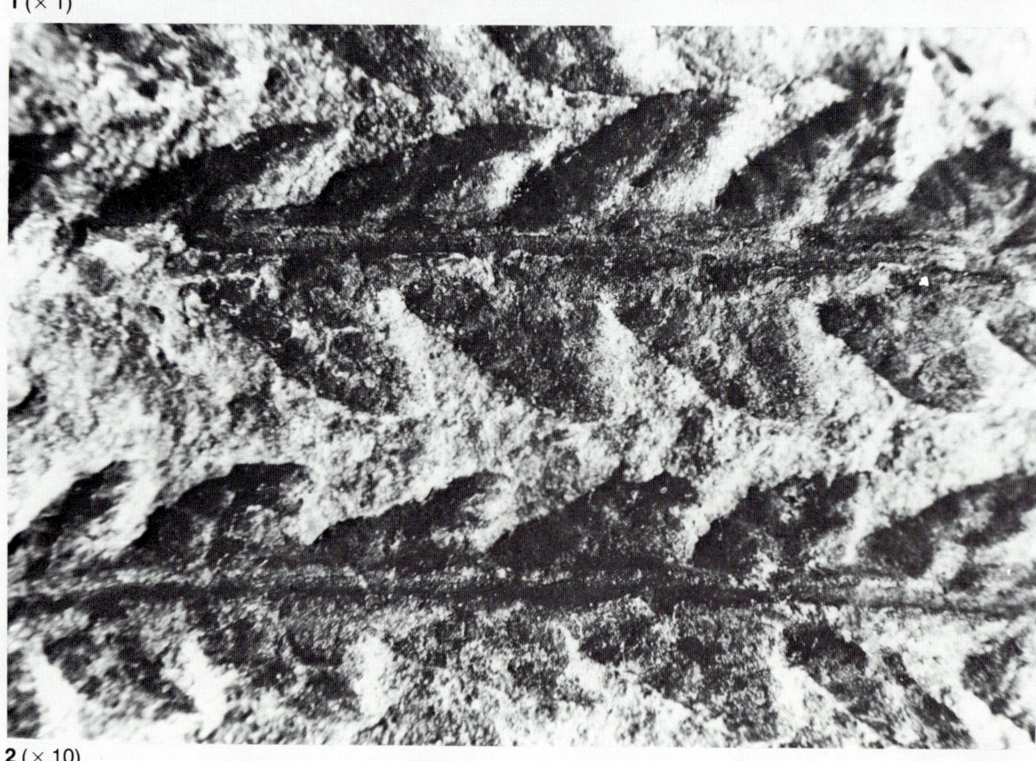

2 (× 10)

1 (× 1.3)

2 (× 3)

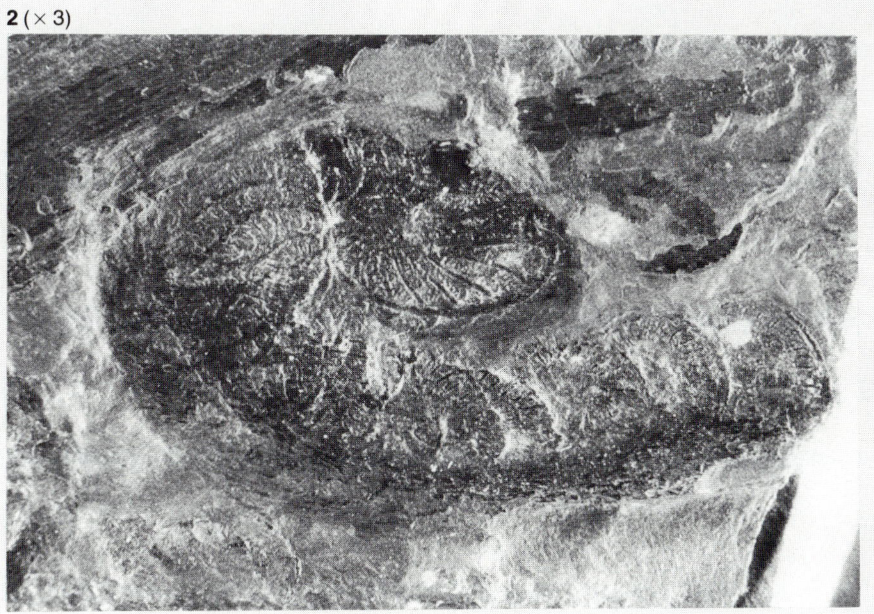

Plate 81

Fig. 1 / *Pecopteris* sp. indet., on display in the Nova Scotia Museum, Halifax, N.S. Unfolding fern frond. From the Morien series, Cape Breton Island (p. 61).

Fig. 2 / Detail of Fig. 1; infrared reflection image of a "fiddlehead," the first on the right in Fig. 1.

Plate 82

Fig. 1 / *Ptychocarpus unitus* (Brongniart) forma *emarginatus* (F-608). Stubbart seam, Prince Mine (p. 62); a slab with many detached pinnae, some of which are detailed on this and Plate 83.

Fig. 2 / Detail of (F-608) showing venation scheme of apical parts of two pinnae; infrared reflection image.

Fig. 3 / *Ptychocarpus unitus* (Brongniart) forma *emarginatus* (F-612). Same location as in Fig. 1. Detail of upper part of a rather "pointed" pinna (note the venation scheme). Infrared reflection image.

Fig. 4 / *Ptychocarpus unitus* (Brongniart) from the famous Mazon Creek flora, Illinois (Pennsylvanian age); center part of pinna, showing confluence of pinnules.

1 (× 0.5)

2 (× 3)

3 (× 3)

4 (× 2)

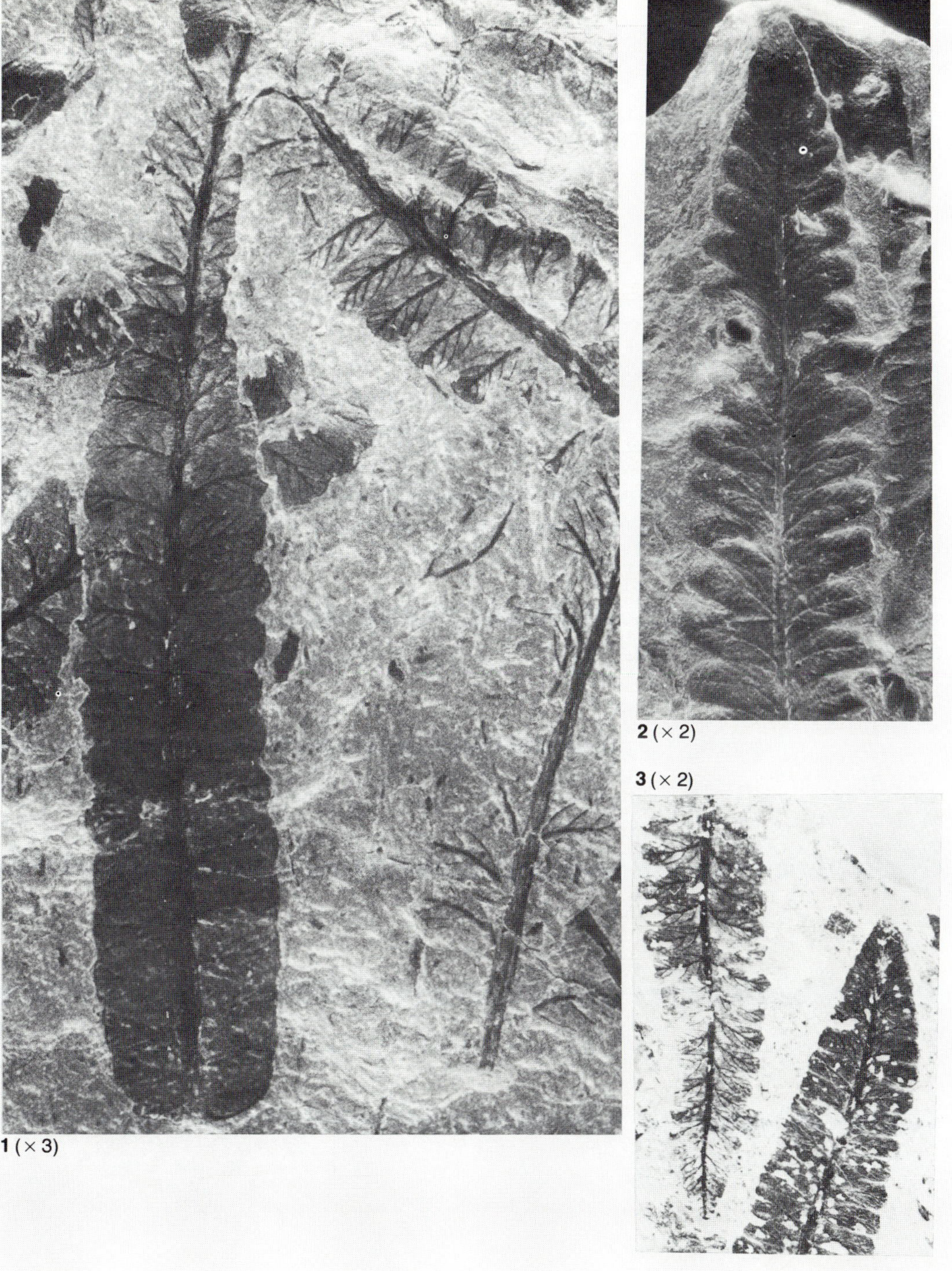

Plate 83

Fig. 1 / *Ptychocarpus unitus* (Brongniart) forma *emarginatus* (F-608). Detail; Stubbart seam, Prince Mine (p. 62); there are six pairs of vascular laterals arising from the midrib of one pinnule in the specimen at the upper right hand corner. Infrared reflection image.

Fig. 2 / *Ptychocarpus unitus* (Brongniart) (F-487). Harbour seam, Lingan Mine (p. 61) note the uncrowded condition of the pinnules on the pinna and compare with other figures. Infrared reflection image.

Fig. 3 / *Ptychocarpus unitus* (Brongniart) forma *emarginatus* (F-611). Stubbart seam, Prince Mine; tips of two pinnae.

Plate 84

Fig. 1 / *Ptychocarpus unitus* (Brongniart) (F-393). Harbour seam, Lingan Mine (p. 61); terminal parts of a frond. Lobation is complete in that the pinnules have merged.

Fig. 2 / *Ptychocarpus* cf. *unitus* (Brongniart). Detail of (F-584) in Fig. 3, showing arrangement of synangia about the midrib; infrared reflection image.

Fig. 3 / *Ptychocarpus* cf. *unitus* (Brongniart) (F-584). Emery seam, Glace Bay; fertile pinna. See Fig. 2 above for detail.

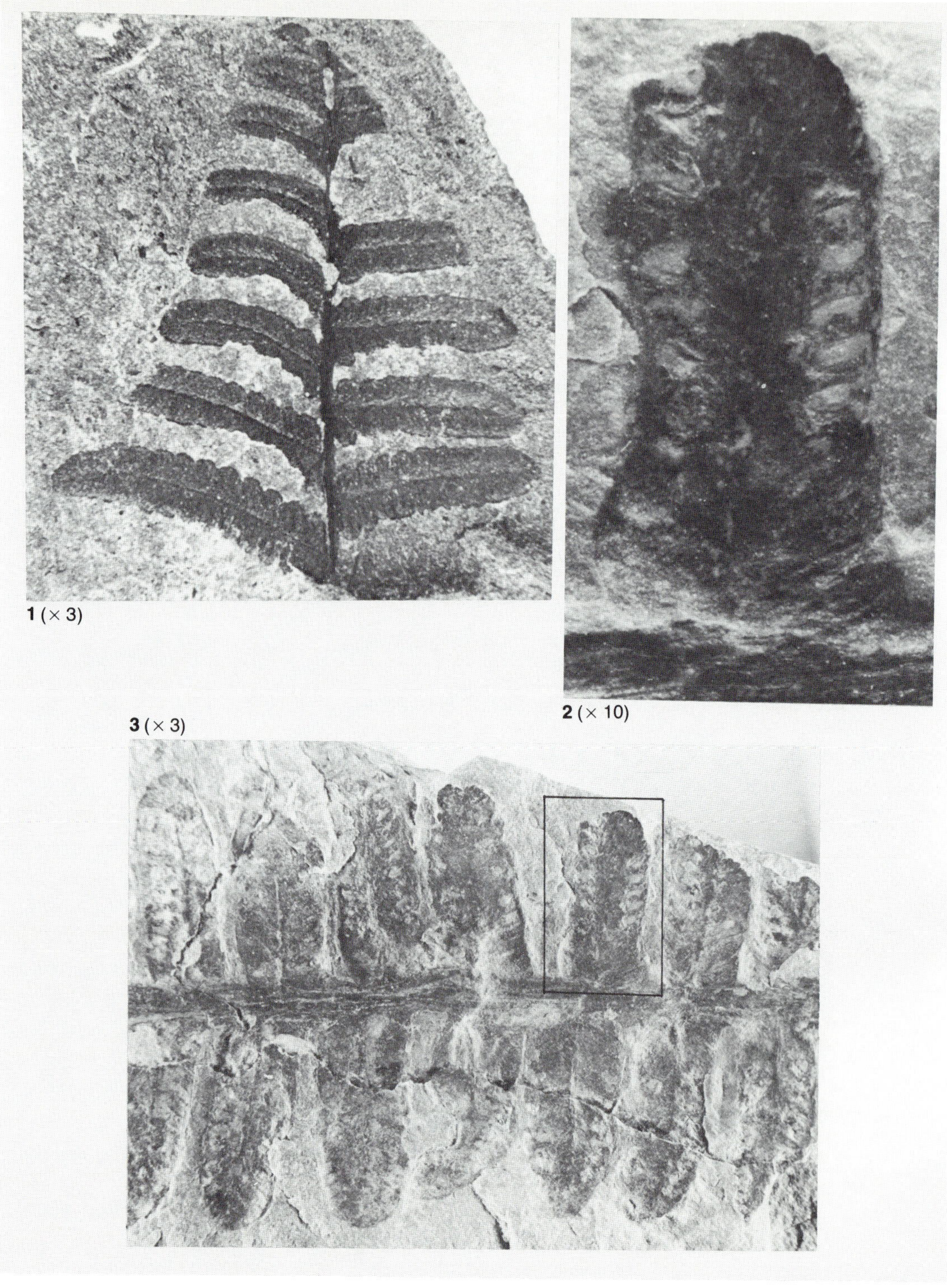

1 (× 13)

2 (× 3)

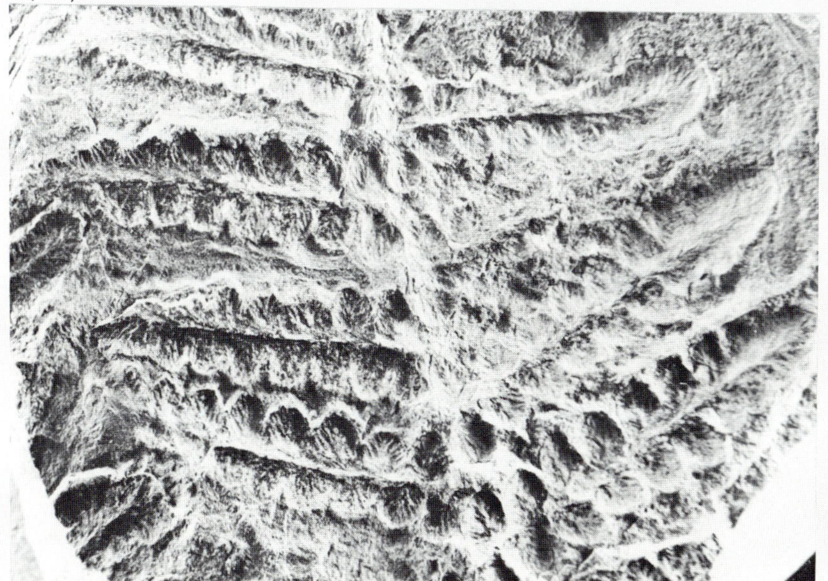

Plate 85

Fig. 1 / *Ptychocarpus unitus* (Brongniart) forma *emarginatus* (F-402-1). Unknown coal measure at Point Aconi (p. 62); detail of two confluent pinnules with stout midribs and rachis (rachis is along bottom edge of photograph) and characteristic venation.

Fig. 2 / *Senftenbergia* sp. indet. (F-173). Emery seam, Glace Bay (p. 64); portion of a frond.

Plate 86

Fig. 1 / *Oligocarpia* cf. *missouriensis* D. White (F-174). Emery seam, Glace Bay (p. 65); partial frond.

Fig. 2 / *Oligocarpia missouriensis* D. White (F-562). Emery seam, Glace Bay; partial frond.

1 (× 5)

2 (× 1)

1 (× 2)

2 (× 3)

Plate 87

Fig. 1 / *Oligocarpia* cf. *missouriensis* D. White (919G1.5). Morien series (p. 65); portion of a frond (infrared reflection image).

Fig. 2 / *Sphenopteris (Oligocarpia?) crenatodentata* Bell (F-440). Harbour seam, Lingan Mine (p. 65); parts of a frond.

Plate 88

Fig. 1 / *Sphenopteris (Oligocarpia?) crenatodentata* Bell (F-191). Phalen seam, Glace Bay (p. 65); detail from an incomplete frond.

Fig. 2 / *Alethopteris* sp. indet. (F-326). Harbour seam, Lingan Mine (p. 66); perhaps a crushed seed case borne by an alethopterid (infrared reflection image).

Fig. 3 / Same specimen as in Fig. 2, photographed on monochromatic film.

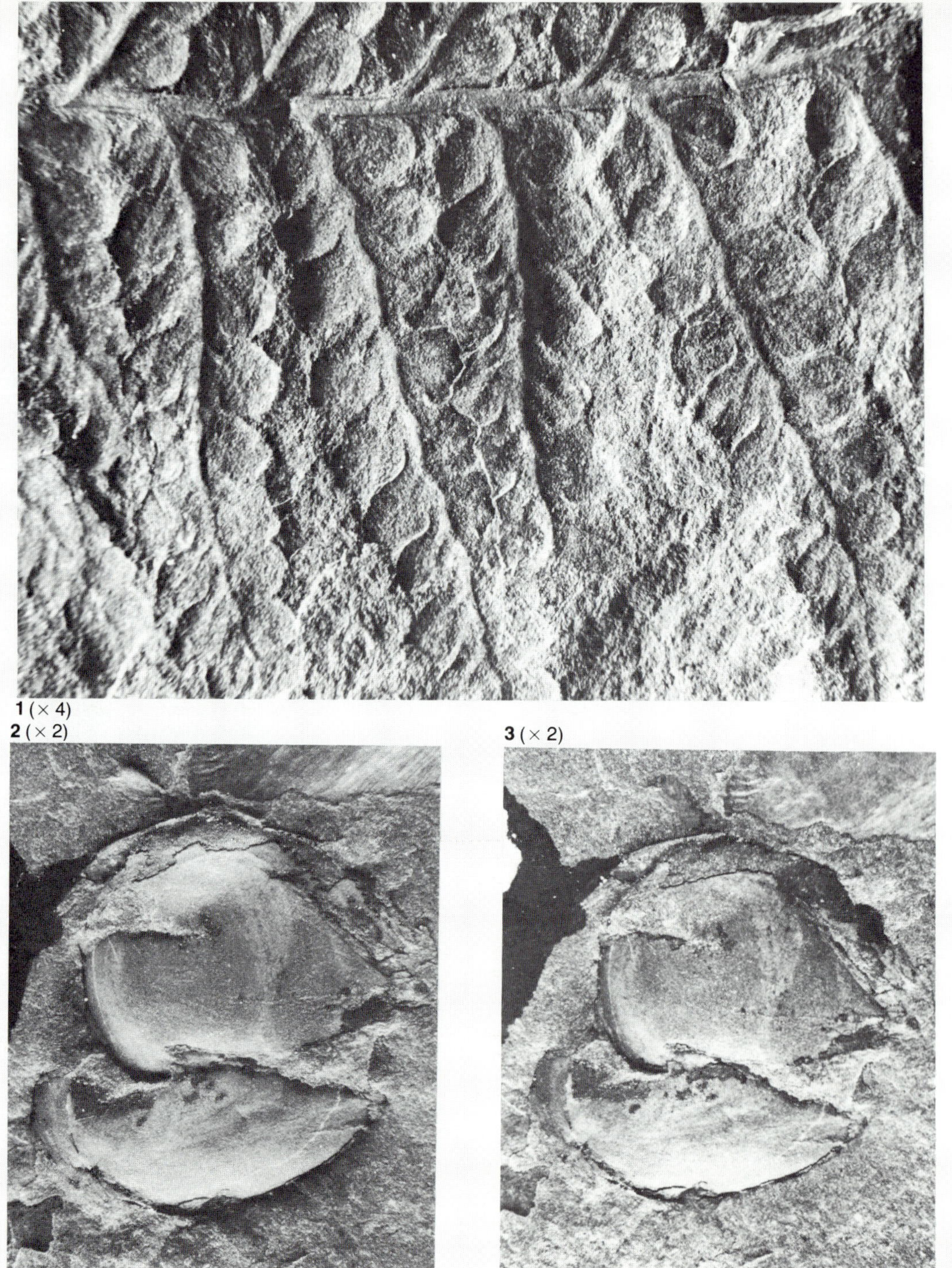

2 (× 3)

1 (× 6)

Plate 89

Fig. 1 / *Sphenopteris* sp. indet. (F-290). Harbour seam, Lingan Mine (p. 66); a fertile portion of a pinna which may be placed in the genus *Oligocarpia* (Abbott, 1954); see also Table 2. Infrared reflection image.

Fig. 2 / *Dicksonites pluckeneti* (Schlotheim) (967G123.17) Morien series (p. 67); detail of a partial frond showing the lowest pinnae, right-hand side of the specimen.

Plate 90

Fig. 1 / *Sphenopteris* cf. *hoeninghausi* Brongniart (F-190). Phalen seam (p. 67); a partial frond.

Fig. 2 / Detail of (F-190) showing venation and shape of foliage; infrared reflection image.

1 (× 2)

2 (× 5)

Plate 91

Fig. 1 / *Annularia radiata* (Brongniart) (F-24). Harbour seam, #12 Mine (p. 68).

Fig. 2 / *Annularia radiata* (Brongniart) (967G10.37). Morien series, Cape Breton Island.

Fig. 3 / *Annularia radiata* (Brongniart) (F-43). Same location as in Fig. 1 above.

Plate 92

Fig. 1 / *Annularia sphenophylloides* (Zenker) (F-601). Point Aconi, unknown coal measure (p. 69).

Fig. 2 / *Annularia sphenophylloides* (Zenker) (F-488). Harbour seam, Lingan Mine.

Both figures are details of specimen whorls.

1 (× 8)

2 (× 7)

1 (× 19)

Plate 93

Fig. 1 / *Annularia sphenophylloides* (Zenker) (F-442). Harbour seam, Lingan Mine (p. 69); a pointed leaf; mucronate structure is clearly visible. (Refer to Gothan and Weyland (1973, p. 195, Abb. 139c) for further details.)

Plate 94

Fig. 1 / *Annularia stellata* (Schlotheim) (F-137). Emery seam (p. 69).

Fig. 2 / *Annularia stellata* (Schlotheim) (F-40). Harbour seam, #12 Mine dump.

1 (× 2)

2 (× 1)

Plate 95

Fig. 1 / *Annularia stellata* (Schlotheim) (967G10.49), Morien series, Cape Breton Island (p. 69).

Fig. 2 / *Annularia stellata* (Schlotheim) from the famous Mazon Creek flora in Illinois (Pennsylvanian age). This specimen should be compared to Plate 96.

Plate 96

Fig. 1 / *Annularia mucronata* Schenk (F-513). Emery seam (p. 68). The main branch is identified by a single arrow; mucronate structure and spatulate leaves are plainly visible at the double arrow.

1 (× 0.9)

1 (× 14)

2 (× 14)

Plate 97

Fig. 1 / *Annularia mucronata* Schenk (F-515). Detail showing mucronate structure (p. 68).

Fig. 2 / *Annularia mucronata* Schenk (F-515). Detail showing dichotomizing mucronate structure.

Plate 98

Fig. 1 / *Asterophyllites equisetiformis* (Sternberg) (F-483). Harbour seam, Lingan Mine (p. 70); foliage attached to a calamarian stem.

1 (× 1)

1 (× 1.5)

2 (× 3)

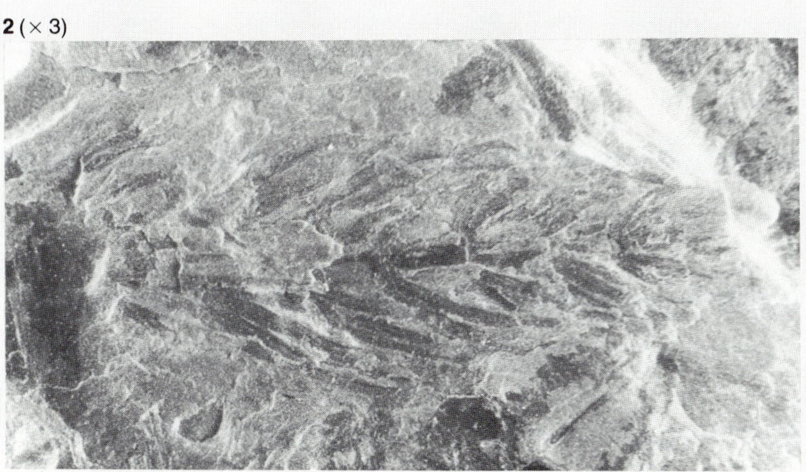

Plate 99

Fig. 1 / *Asterophyllites equisetiformis* (Sternberg) (F-119). Emery seam (p. 70); note attachment of foliar branches to calamarian stem.

Fig. 2 / *Asterophyllites equisetiformis* (Sternberg) (F-248). Stubbart seam, Prince Mine, Point Aconi; infrared reflection image.

Plate 100

Fig. 1 / *Calamites carinatus* Sternberg (F-245). Stubbart seam, Point Aconi, Prince Mine (p. 71).

Fig. 2 / *Calamites cisti* Brongniart (967G10.9), Morien series, Cape Breton Island (p. 71).

1 (× 0.7)

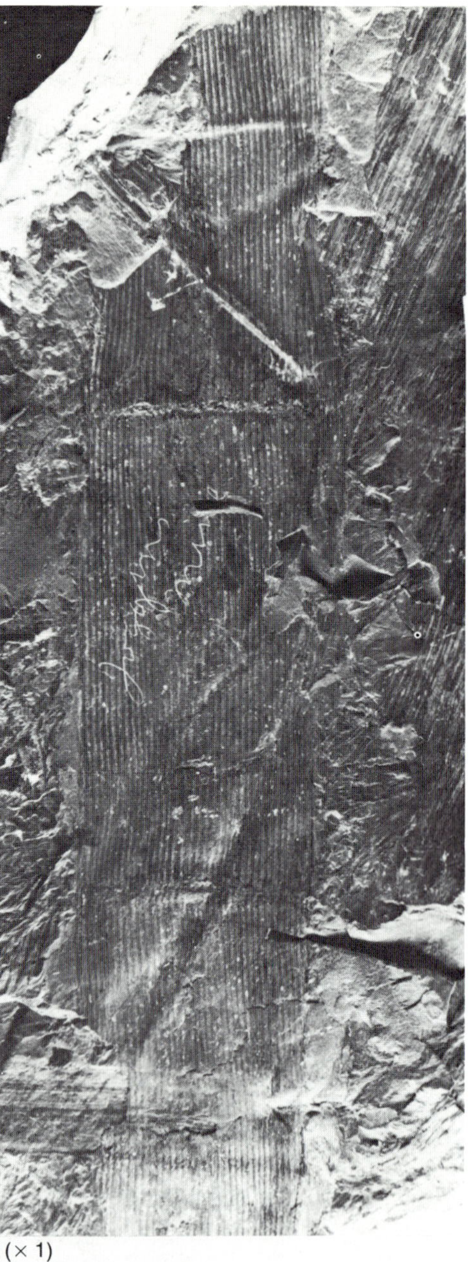

2 (× 1)

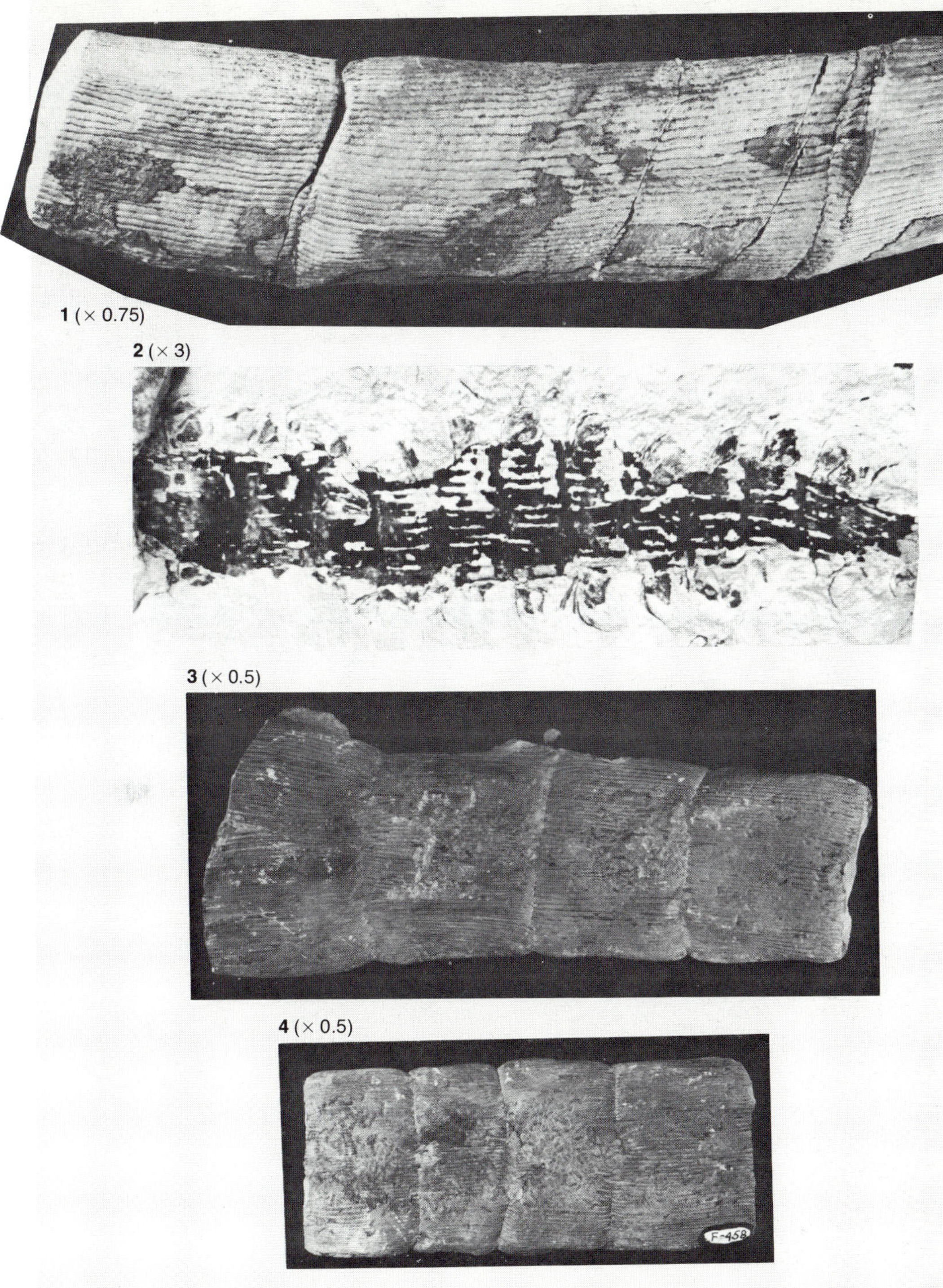

Plate 101

Fig. 1 / *Calamites cistiiformis* Stur (F-180). Phalen seam (p. 71).

Fig. 2 / *Calamites* species (F-252). Stubbart seam, Point Aconi, Prince Mine (p. 72); fructification of a calamite.

Fig. 3 / *Calamites multiramis* Weiss (F-458). Emery seam, Glace Bay (p. 72).

Fig. 4 / Continuation of the specimen in Fig. 3.

Plate 102

Fig. 1 / *Calamites ramosus* Artis (901G3.1). One and three-quarter miles from the mouth of Halfway River, Hants Co., N.S. (p. 72); scar of branch attachment.

Fig. 2 / *Calamites ramosus* Artis (967G143.39). McLeod Brook, one mile southwest of Westville, Pictou County, Pictou series.

2 (× 1)

1 (× 3)

1 (× 1)

2 (× 1)

Plate 103

Fig. 1 / *Calamites* sp. group ? *Calamites varians* Sternberg (976GF29.1). Unknown locality (p. 73); note scars where branches were attached.

Fig. 2 / *Calamites suckowi* Brongniart (967G30.1). Wallace Quarry, Cumberland Co., N.S. Collected by the Very Rev. D. Crawford from stones used in the construction of All Saints Cathedral, Halifax, N.S. (p. 73). A branch of a calamite.

Plate 104

Fig. 1 / *Calamites suckowi* Brongniart (F-181). Phalen seam, Glace Bay (p. 73); detail of nodes and xylem bundles.

Fig. 2 / *Calamites suckowi* Brongniart (967G143.25). Harbour seam, Sydney Mines.

1 (× 5)

2 (× 1)

1 (× 1)

2 (× 1)

Plate 105

Fig. 1 / *Calamites suckowi* Brongniart (F-198). Unspecified coal measure on Cape Breton Island (p. 73); a pith cast.

Fig. 2 / *Calamites suckowi* Brongniart (967G18.1). Watering Brook, West Bay, Cumberland Co., N.S.

Plate 106

Fig. 1 / *Calamites suckowi* Brongniart (F-497). Joggins, Cumberland Co., N.S. (p. 73); shown are nodes and xylem bundles.

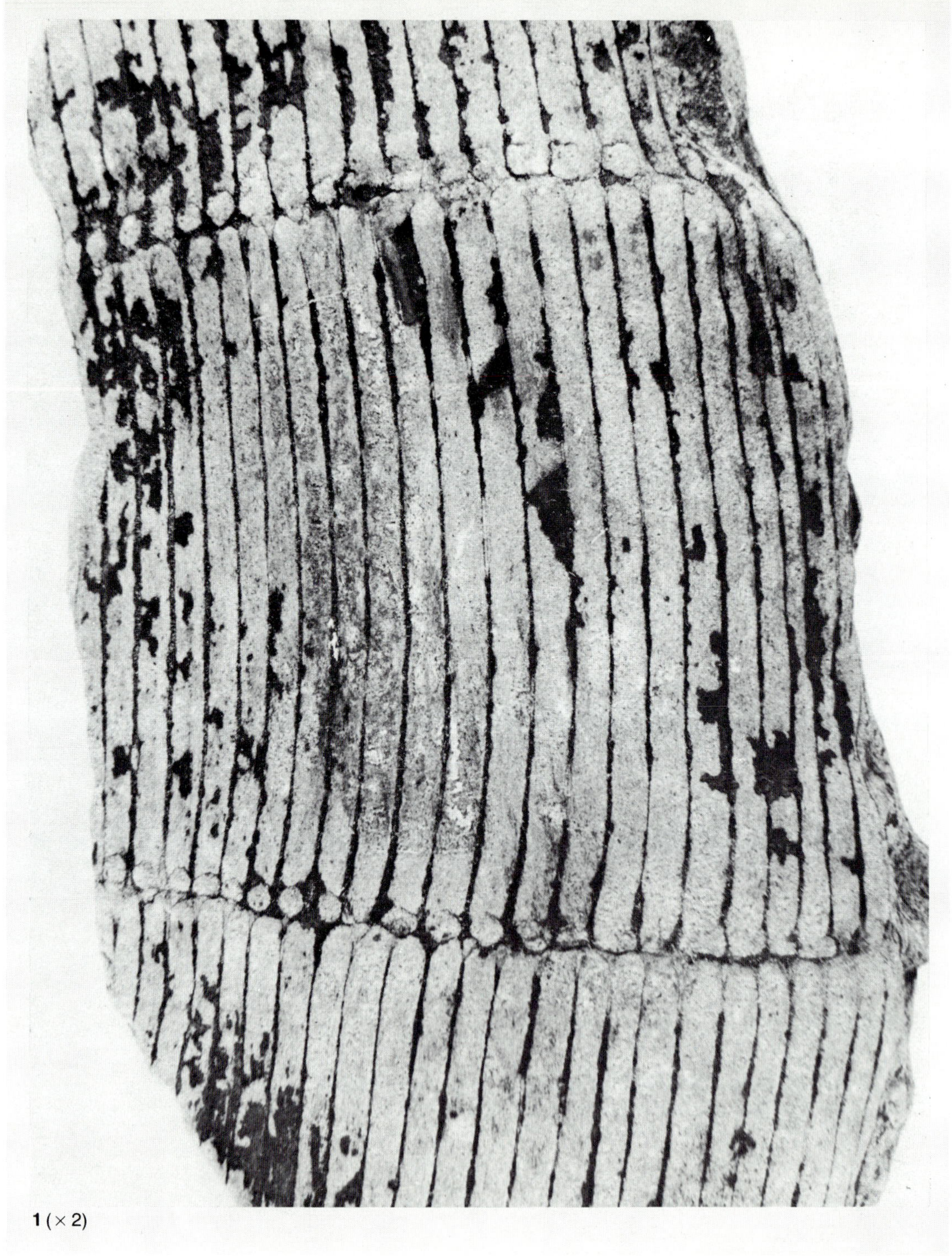

1 (× 2)

1 (× 10)

2 (× 1)

3 (× 3)

Plate 107

Fig. 1 / *Sphenophyllum cuneifolium* (Sternberg) (F-23). Harbour seam, #12 Mine dump (p. 75); detail of a wedge-shaped leaflet (to which the name refers) from a complete whorl.

Fig. 2 / *Calamites waldenburgensis* Kidston (967G10.41). Morien series, Cape Breton Island (p. 74).

Fig. 3 / *Macrostachya infundibuliformis* (Brongniart) (F-259). Stubbart seam, Prince Mine (p. 74); fragment of a cone.

Plate 108

Fig. 1 / *Macrostachya infundibuliformis* (Brongniart) (F-647). Stubbart seam, Prince Mine (p. 74); a complete cone attached to an axis (attachment is difficult to trace on the photograph).

Fig. 2 / *Palaeostachya* sp. indet. (F-604). Stubbart seam, Prince Mine (p. 74).

Fig. 3 / *Calamites waldenburgensis* Kidston (967G11.2). Shore at McCarren's Brook, South Joggins, Cumberland Co., N.S. (p. 74).

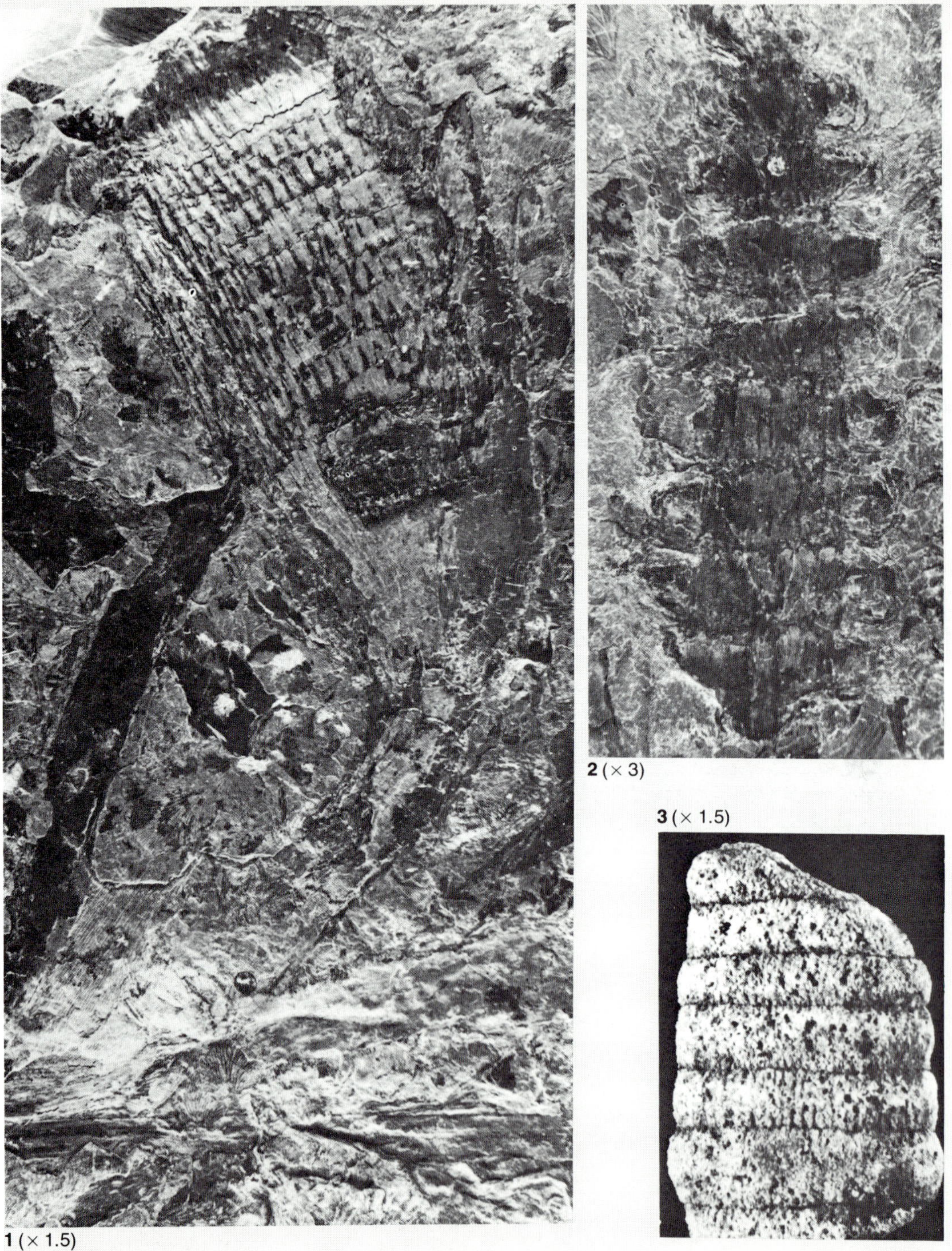

1 (× 2)

2 (× 2)

3 (× 3)

Plate 109

Fig. 1 / *Sphenophyllum cuneifolium* (Sternberg) (F-196). Phalen seam (p. 75).

Fig. 2 / *Sphenophyllum cuneifolium* (Sternberg) (F-53). Mc Aulay seam.

Fig. 3 / *Sphenophyllum cuneifolium* (Sternberg) (F-45). Harbour seam at #12 Mine dump.

Plate 110

Fig. 1 / *Sphenophyllum emarginatum* (Brongniart) (F-18). Harbour seam, #12 Mine dump (p. 75).

Fig. 2 / *Sphenophyllum emarginatum* (Brongniart) (F-46). Same location as above.

Fig. 3 / *Sphenophyllum emarginatum* (Brongniart) (967G10.43). Unspecified coal measure on Cape Breton Island.

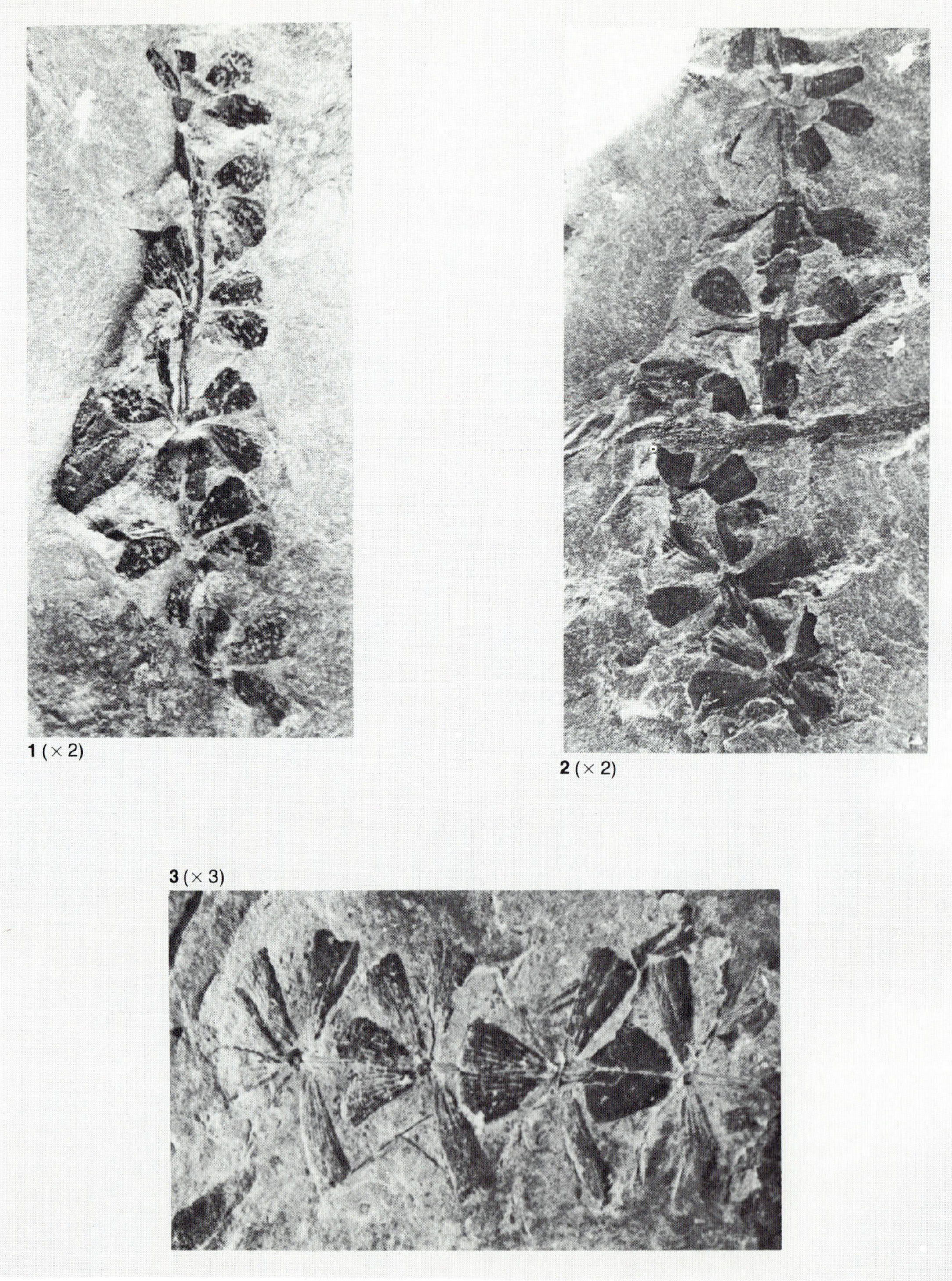

1 (× 2)

2 (× 2)

3 (× 3)

Plate 111

Fig. 1 / *Sphenophyllum majus* (Bronn) (F-298). Harbour seam, Lingan Mine (p. 77).

Fig. 2 / *Sphenophyllum majus* (Bronn) (F-297). Same location as above.

Fig. 3 / *Sphenophyllum majus* (Bronn) (F-302). Same location as above.

Plate 112

Fig. 1 / *Sphenophyllum oblongifolium* (Germar and Kaulfuss) (F-370). Harbour seam, Lingan Mine (p. 78).

Fig. 2 / *Sphenophyllum oblongifolium* (Germar and Kaulfuss) (F-481). Phalen seam, Glace Bay.

Fig. 3 / *Sphenophyllum oblongifolium* (Germar and Kaulfuss) (967G10.54). Morien series, Cape Breton Island.

1 (× 2)

2 (× 2)

3 (× 3)

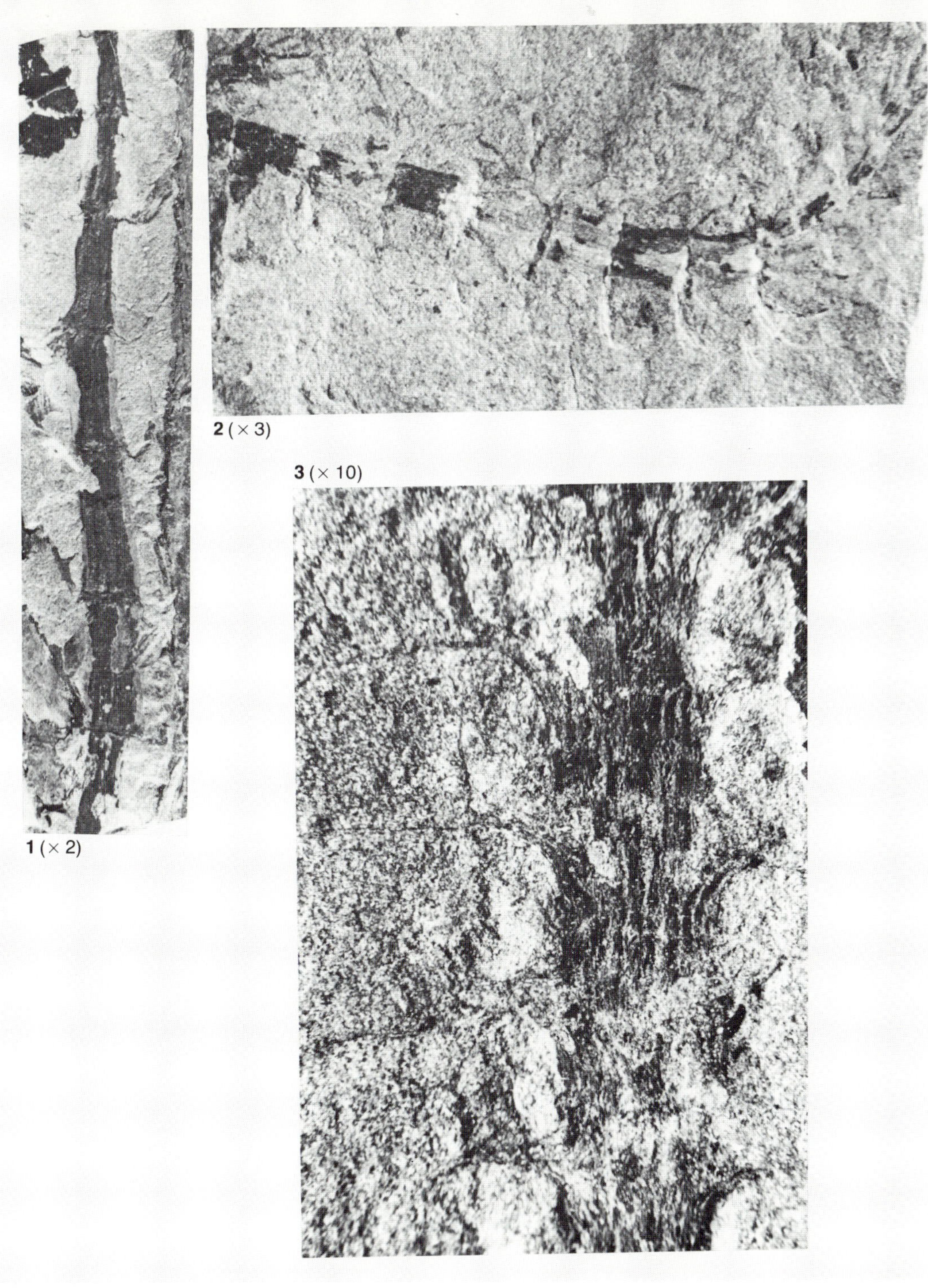

Plate 113

Fig. 1 / *Sphenophyllum* sp. indet. (F-301). Harbour seam, Lingan Mine (p. 78); a stem.

Fig. 2 / *Sphenophyllum* sp. indet. (F-88). Shoemaker seam (p. 78); fertile.

Fig. 3 / *Sphenophyllum* sp. indet. (F-91). Shoemaker seam (p. 78); detail of specimen showing possible spore outlines (the light ovate areas nearest the axis).

Plate 114

Fig. 1 / *Sphenophyllum* cf. *trichomatosum* Stur (F-63). Mc Aulay seam (p. 79); detail of jointed stem with punctae (?) at internodes. The internodes are 3 mm apart. Compare to Bell's figures (1938, p. 90 and Pl. 93, Fig. 8).

Fig. 2 / *Lepidodendron aculeatum* Sternberg (967G10.85). Unspecified coal measure on Cape Breton Island (p. 79).

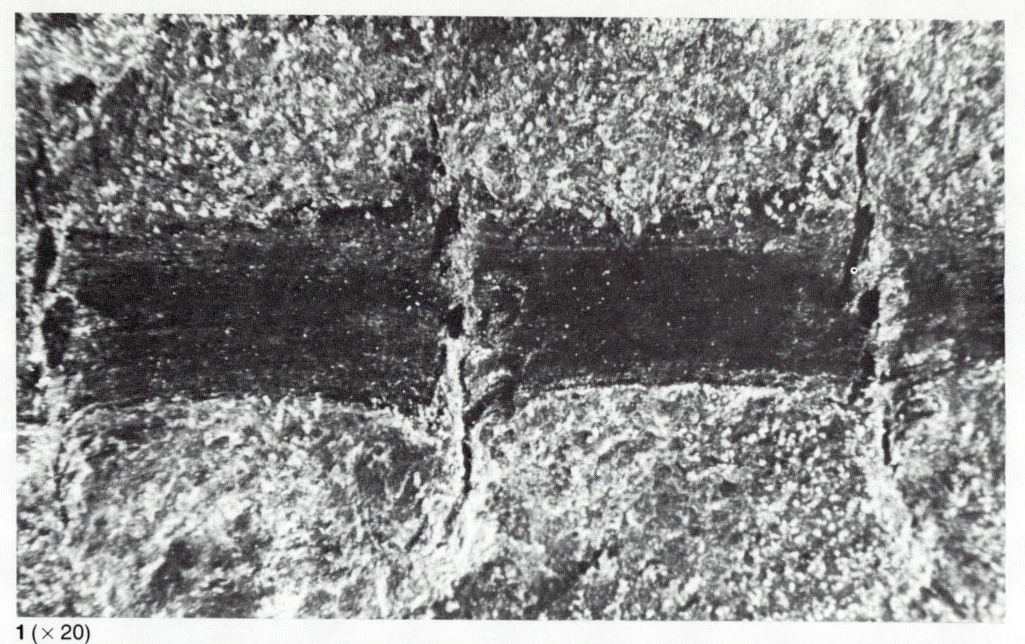

1 (× 20)

2 (× 1)

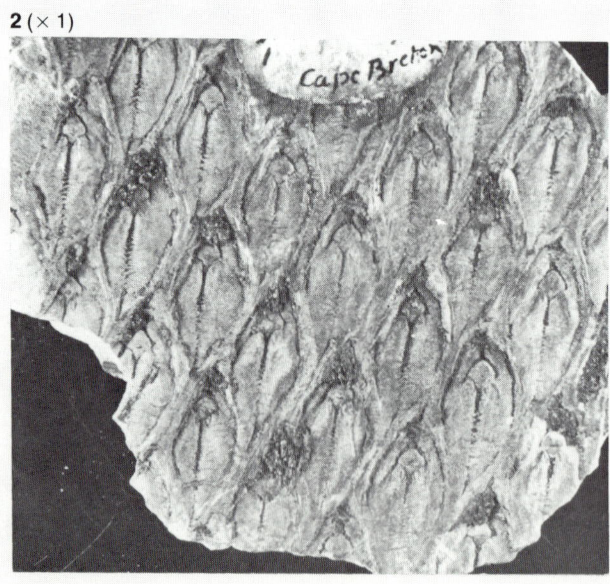

1 (× 2)

2 (× 2)

Plate 115

Fig. 1 / *Lepidodendron aculeatum* Sternberg (F-288). Harbour seam, Lingan Mine (p. 79); leaf cushions with inner structures (parichnos and leaf scars).

Fig. 2 / *Lepidodendron aculeatum* Sternberg (F-288). The coalified layer for the most part covers the leaf cushions of this specimen; infrared reflection image.

Plate 116

Fig. 1 / *Lepidodendron bretonense* Bell (F-29). Harbour seam, #12 Mine dump (p. 80); detail of leaf cushion marked "B", with leaf scar marked "L."

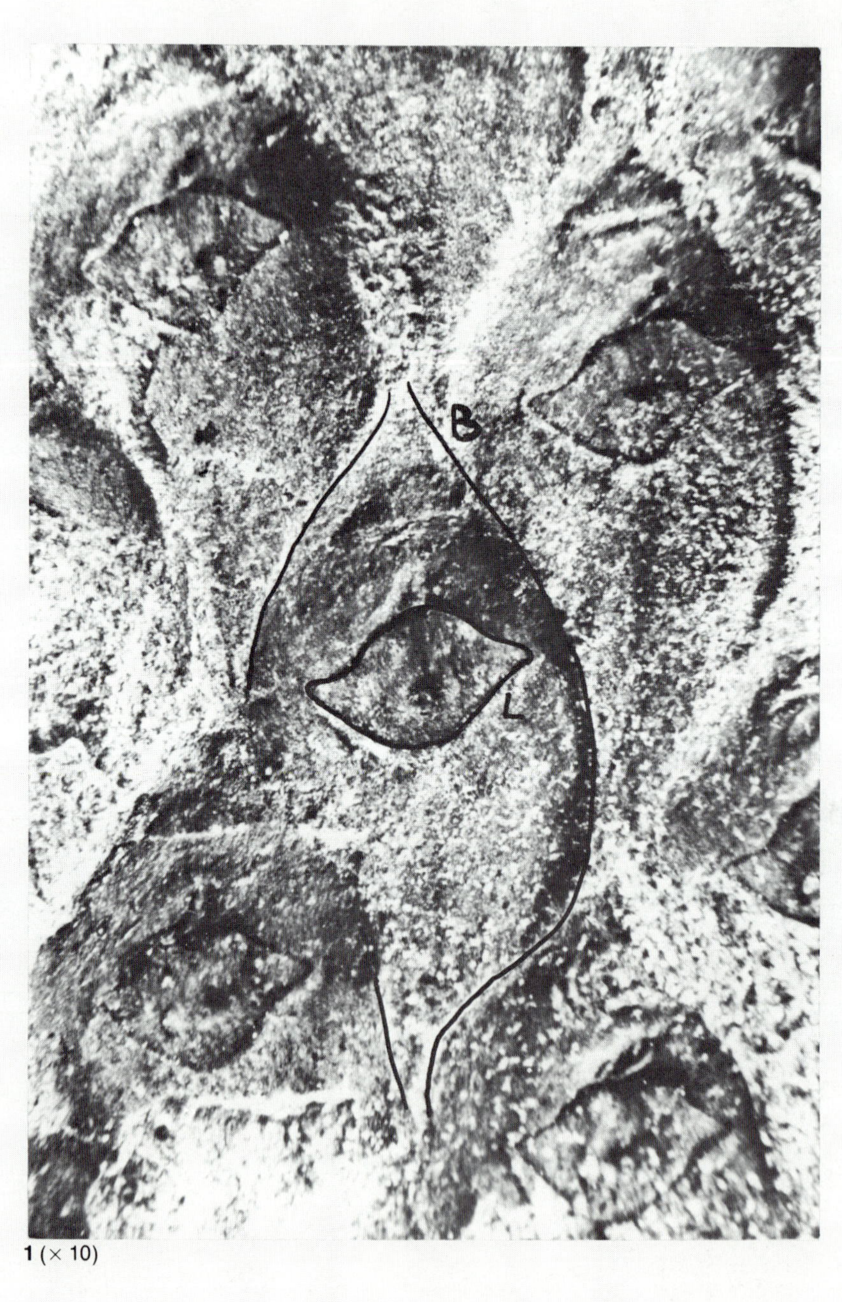

1 (× 10)

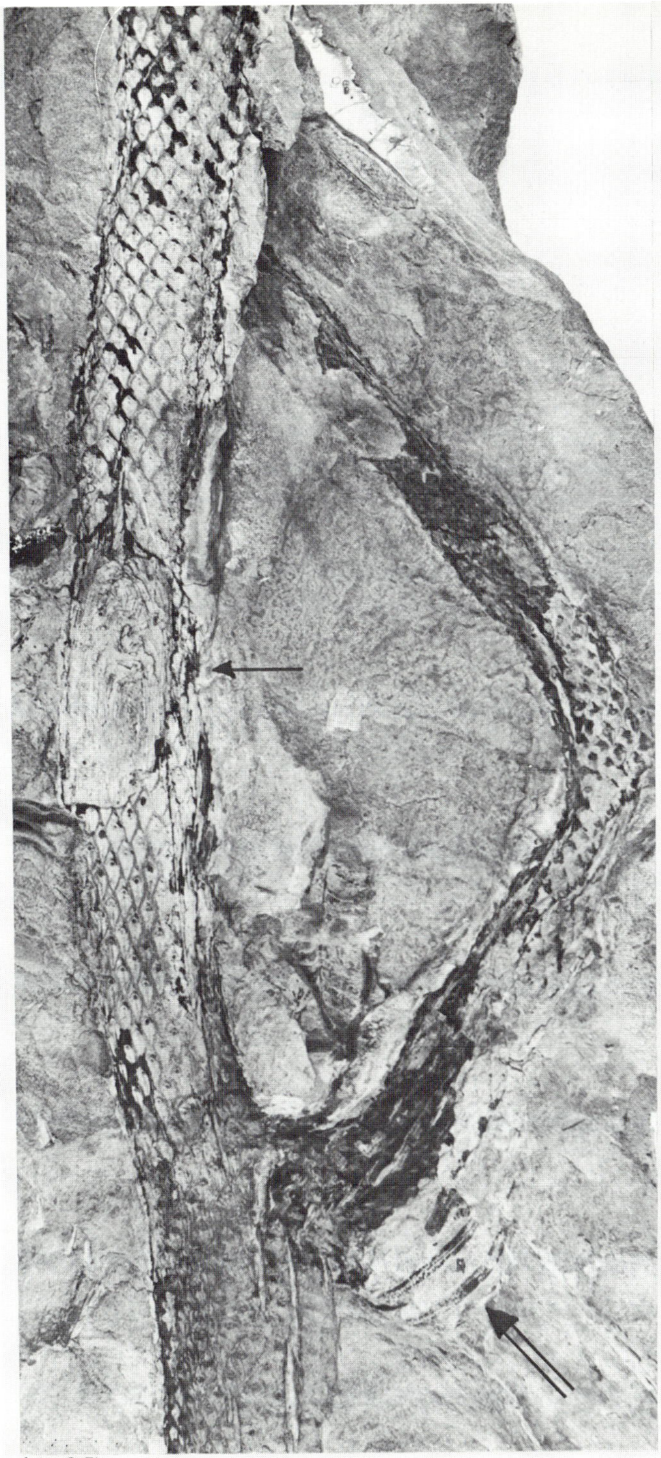

1 (× 0.5)

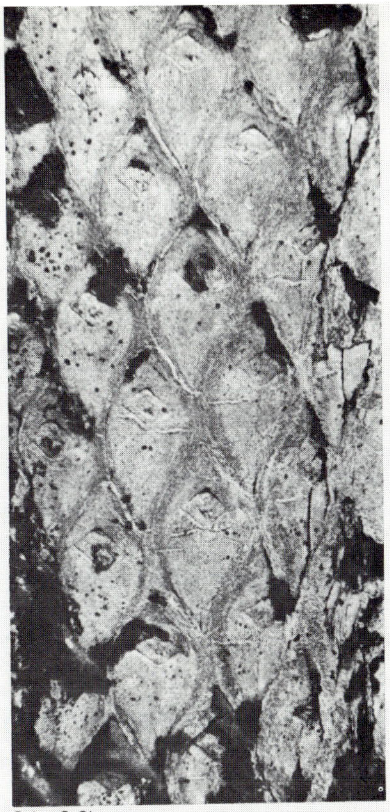

2 (× 2.6)

3 (× 2.6)

Plate 117

Fig. 1 / *Lepidodendron pictoense* Dawson (F-670). Phalen seam, Glace Bay (p. 81); a 45 cm long branching specimen. A single arrow in the photograph identifies a branch scar. Leafy shoots are visible on the lower right of the main stem (double arrow).

Fig. 2 / *Lepidodendron pictoense* Dawson (F-507). Phalen seam, Glace Bay; detail of arrangement and shape of leaf cushions and parichnos scars. The black material in the photograph is residual carbon.

Fig. 3 / *Lepidodendron pictoense* Dawson (F-507). The material which covers leaf cushions and scars shown in Fig. 1.

Figs. 1 and 2 are parts of one fossil tree but their positions relative to one another are not known. F-670 was discovered and collected one year later (August, 1977) than F-507.

Plate 118

Fig. 1 / *Lepidodendron bretonense* Bell (F-656). Stubbart seam, Prince Mine (p. 80); detail of a 13 cm long specimen.

Fig. 2 / *Lepidodendron bretonense* Bell (F-294). Harbour seam, Lingan Mine; infrared reflection image of pyrite replacement in the coalified fossil.

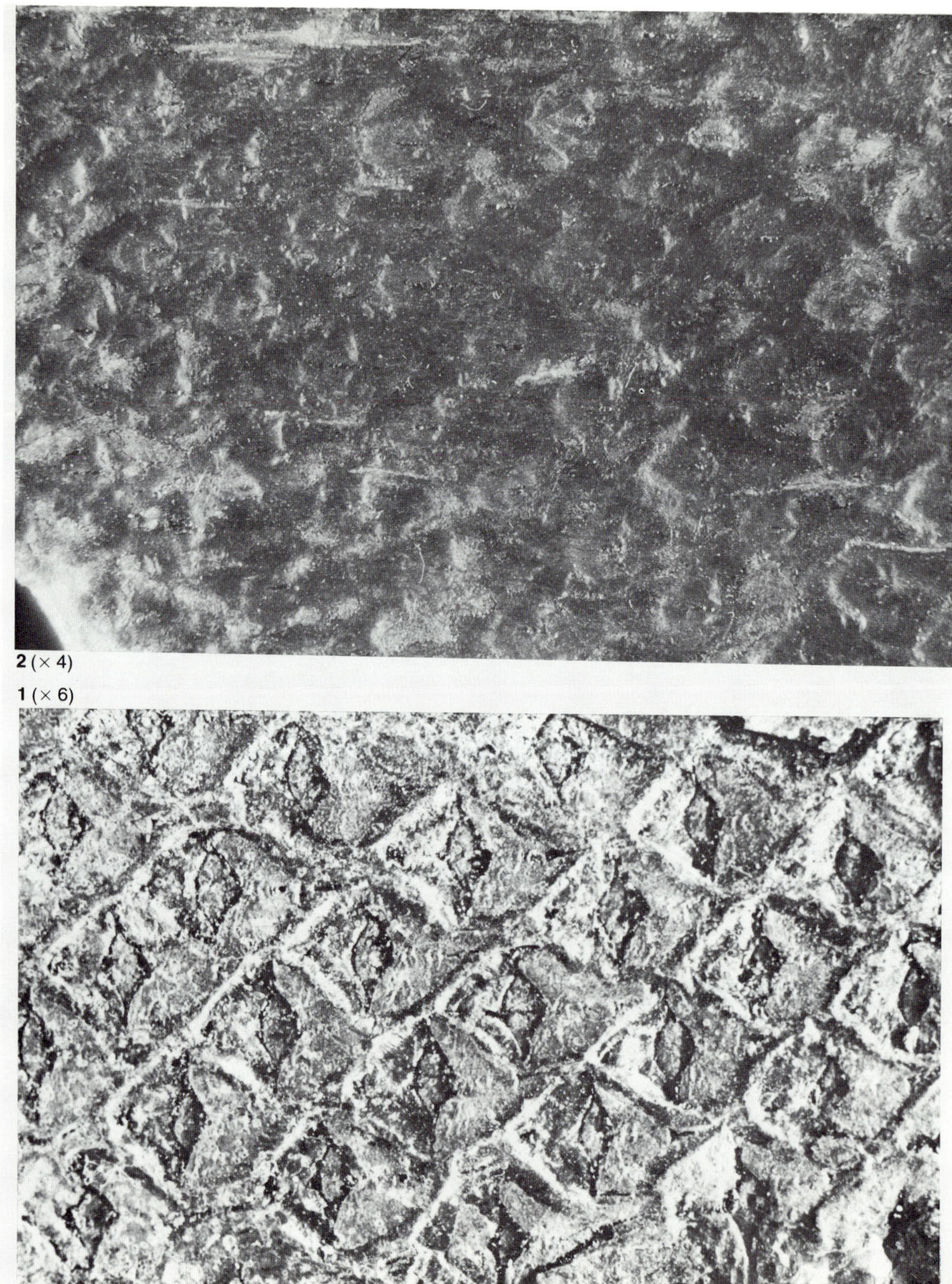

2 (× 4)

1 (× 6)

1 (× 8)

2 (× 3)

3 (× 4)

Plate 119

Fig. 1 / *Lepidodendron dawsoni* Bell (F-299). Harbour seam, #12 Mine dump (p. 80).

Fig. 2 / *Lepidodendron dawsoni* Bell (F-496-1). Harbour seam, Lingan Mine.

Fig. 3 / *Lepidodendron dawsoni* Bell (F-97b). Shoemaker seam. The specific identity of this specimen is uncertain. The light area near its tip is pyrite replacement (infrared reflection image).

Plate 120

Fig. 1/ *Lepidodendron* sp. (F-45-1). Harbour seam, #12 Mine dump (p. 82); leafy stem.

Fig. 2 / *Lepidodendron ophiurus* Brongniart (967G10.44). Morien series, Cape Breton Island (p. 81).

Fig. 3 / *Lepidodendron pictoense* Dawson (F-436). Harbour seam, Lingan Mine (p. 81); note the triangular outline of leaf scars.

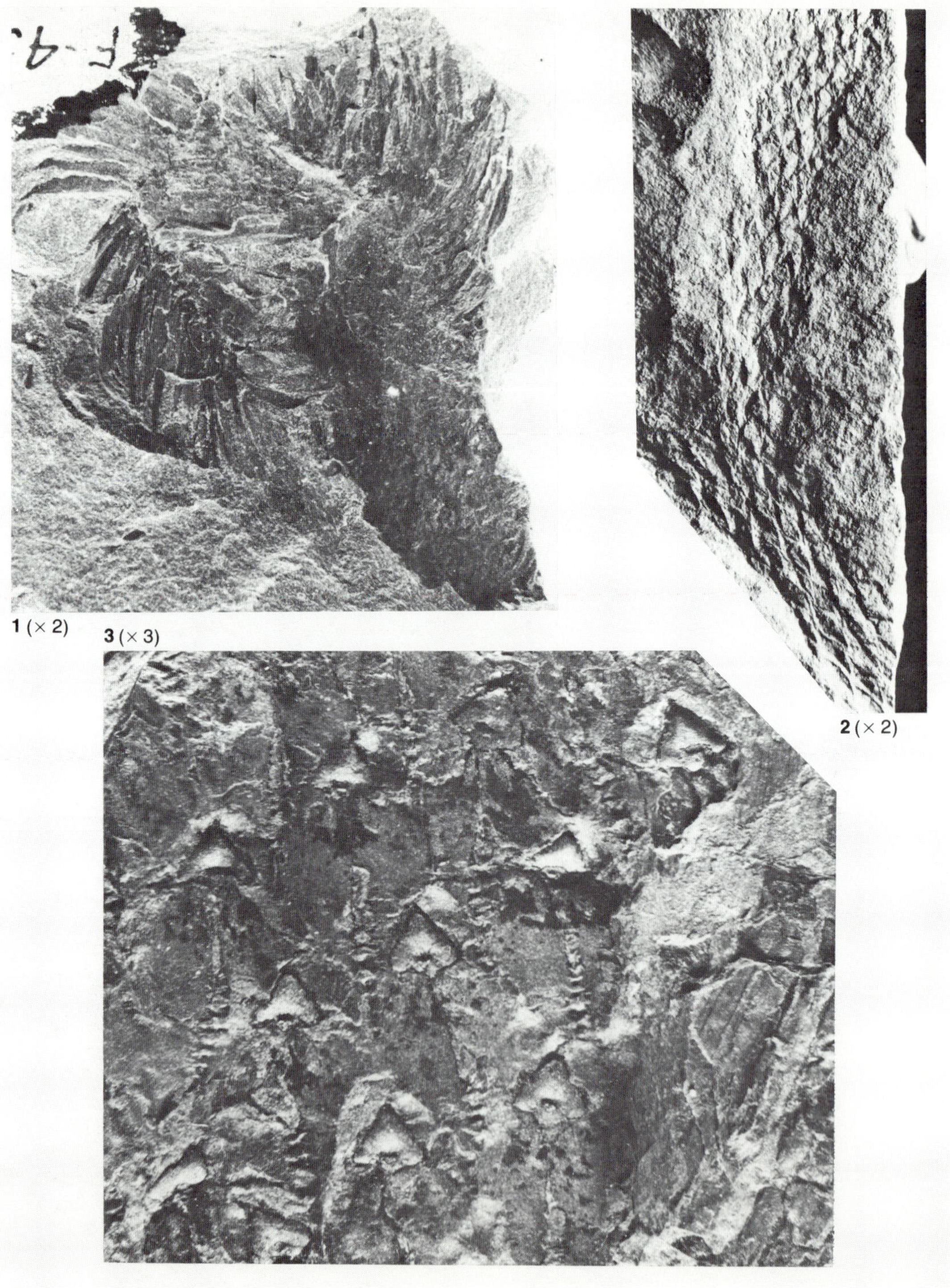

1 (× 2)

2 (× 2)

3 (× 3)

1 (× 2)

2 (× 3)
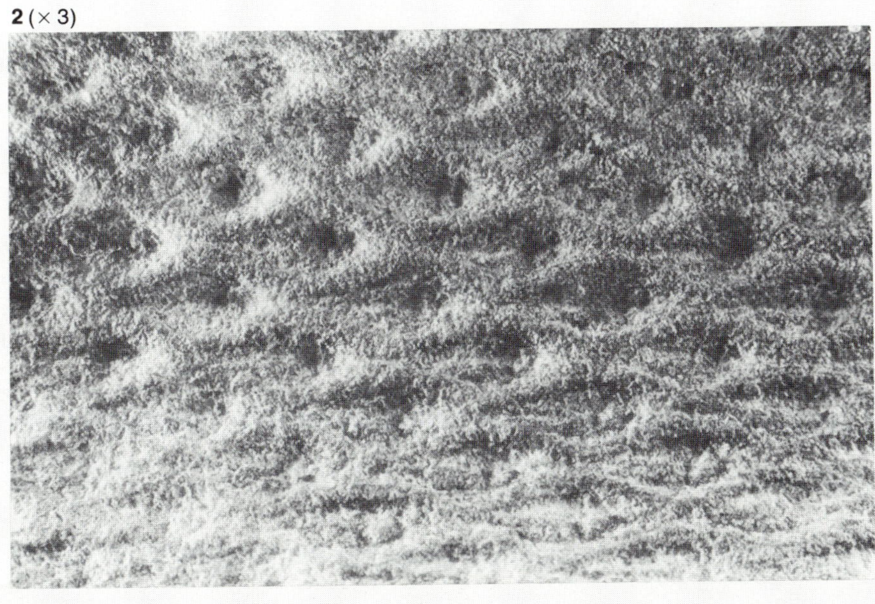

Plate 121

Fig. 1 / *Lepidodendron* sp. indet. (F-354). Harbour seam, Lingan Mine (p. 81).

Fig. 2 / *Lepidodendron* sp. indet. (967G36.11). Unspecified coal seam, Morien series; infrared reflection image of the mold of the fossil.

Plate 122

Fig. 1 / *Lepidodendron* sp. (F-89). Shoemaker seam (p. 82); twigs.

Fig. 2 / *Lepidodendron* sp. (F-467). Same location as above; twigs.

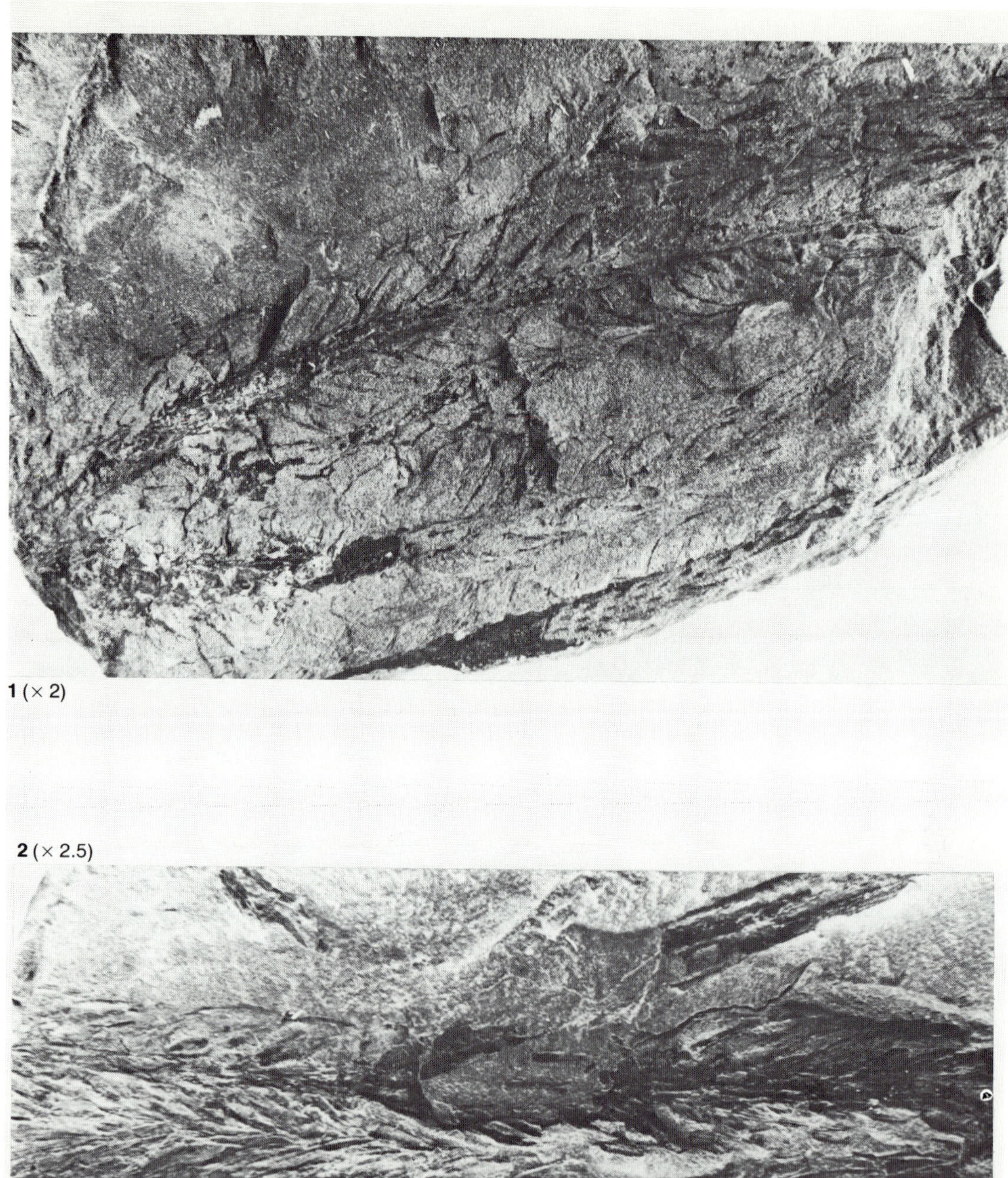

1 (× 2)

2 (× 2.5)

1 (× 1)

2 (× 1)

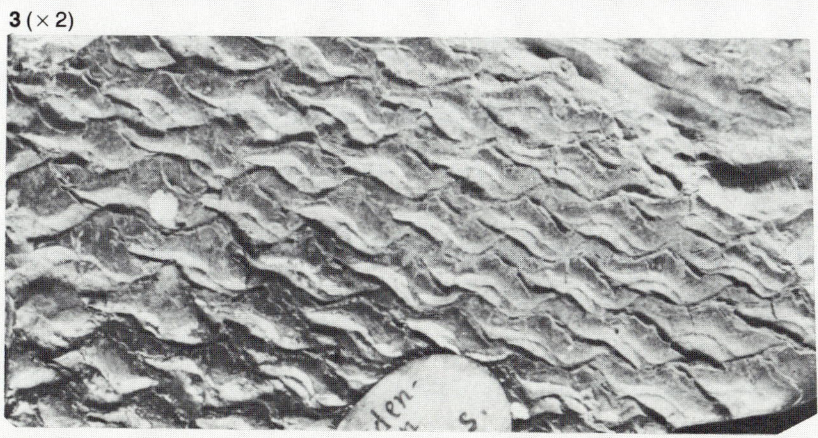

3 (× 2)

Plate 123

Fig. 1 / *Lepidodendropsis corrigatum* (Dawson) (967G35.1). Lower Carboniferous, Horton, N.S. (p. 82).

Fig. 2 / *Lepidophloios laricinus* Sternberg (967G10.15). Morien series, Cape Breton Island, (p. 82).

Fig. 3 / *Lepidophloios laricinus* Sternberg (967G31.4). South Joggins, Cumberland Co., N.S.

Plate 124

Fig. 1 / *Lepidophloios laricinus* Sternberg (967G36.12). Morien series, Cape Breton Island (p. 82); raised leaf scars in spiral sequence, characteristic of the form *Halonia tortuosa*, are clearly visible.

Fig. 2 / *Lepidophyllum* sp. (F-411). Upper Bonar seam, Point Aconi (p. 83).

Fig. 3 / *Lepidostrobophyllum lanceolatum* (Lindley and Hutton) var. *constrictum* Bell (F-506). Phalen seam (p. 84); the constriction on the sporophyll is marked by an arrow.

Fig. 4 / *Lepidostrobophyllum lanceolatum* (Lindley and Hutton) var. *constrictum* Bell (F-506). Another specimen on the F-506 block.

1 (× 1)

2 (× 4)

4 (× 4)

Plate 125

Fig. 1 / *Lepidostrobus* sp. (F-655) with sporophylls of the *Lepidostrobophyllum jenneyi* type. Stubbart seam, Prince Mine (p. 83); note "shedded" sporophylls (for detail see Figs. 3 and 4 below).

Fig. 2 / *Lepidostrobophyllum* cf. *jenneyi* (D. White) (F-483-1). Harbour seam, Lingan Mine (p. 83).

Fig. 3 / Detail of a sporophyll on (F-655) in Fig. 1, where its location is marked "3".

Fig. 4 / Detail of a sporophyll on (F-655) in Fig. 1, where its location is marked "4".

Plate 126

Fig. 1 / *Lepidostrobus* sp. indet. (F-328). Harbour seam, Lingan Mine (p. 85).

Fig. 2 / *Lepidostrobus lanceolatus* Lindley and Hutton (967G10.21). Morien series, Cape Breton Island (p. 84).

Fig. 3 / *Lepidostrobophyllum lanceolatum* (Lindley and Hutton) (967G20.10). Stellarton series (p. 84).

1 (× 2)

2 (× 1)

1 (×10)

2 (× 1)

Plate 127

Fig. 1 / *Lepidostrobus* cf. *mintoensis* Wilson (F-471). Phalen seam (p. 85); detail showing a triangular leaf of the specimen in Fig. 2.

Fig. 2 / *Lepidostrobus* cf. *mintoensis* Wilson (F-471). Phalen seam.

Plate 128

Fig. 1 / *Sigillaria* cf. *brardi* Brongniart (F-350). Harbour seam, Lingan Mine (p. 86).

Fig. 2 / Detail of (F-350) using infrared reflection photography. Leaf cushions are discernible, and a tessellate structure is visible, especially on the top cushions. (See Plate 129).

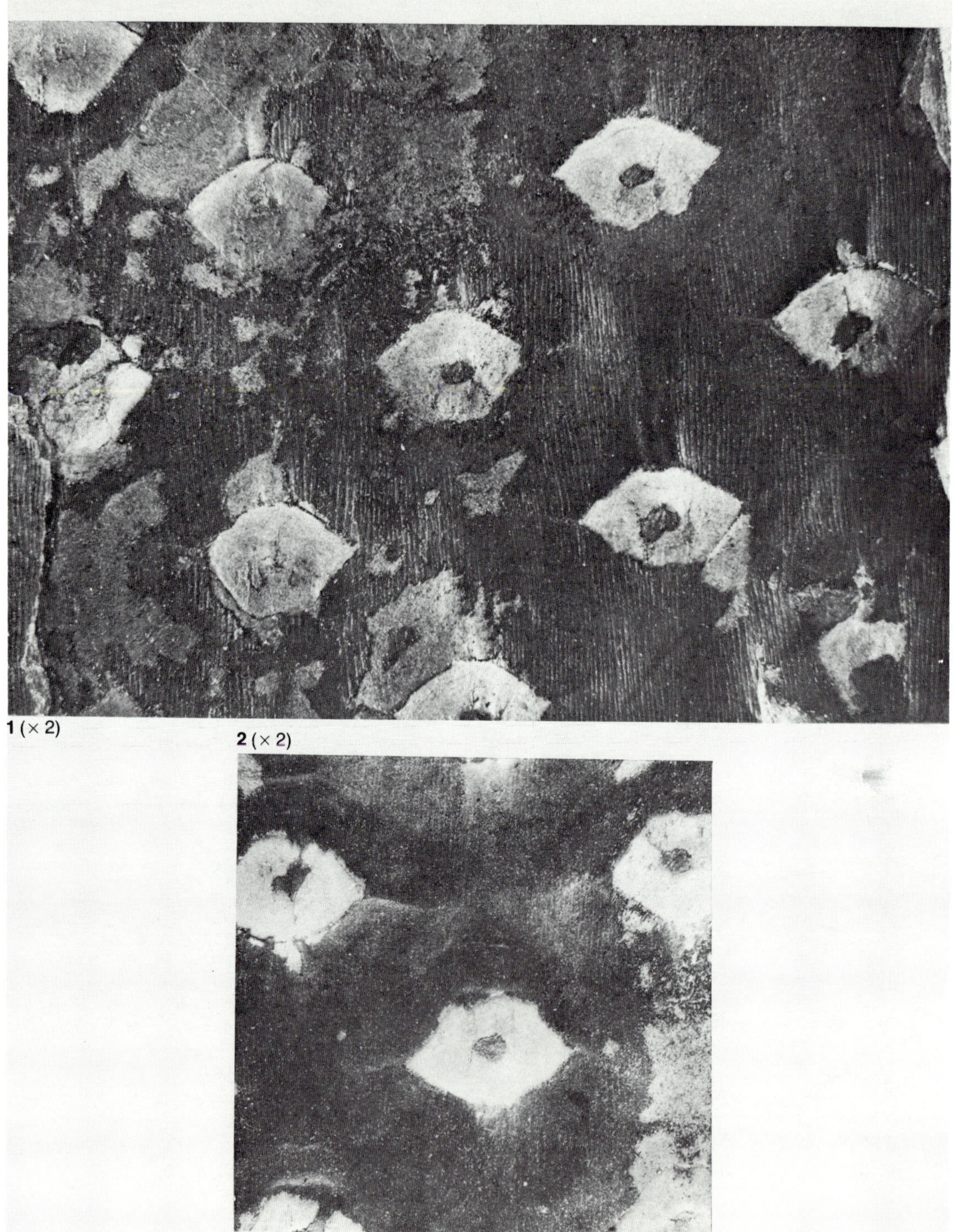

1 (× 2)

2 (× 2)

1 (× 5)

Plate 129

Fig. 1 / *Sigillaria* cf. *brardi* Brongniart (F-350) illustrating visibility of leaf cushions under oblique (partially polarized) white light; note horizontal folds on the hexagonal leaf scars (p. 86).

Plate 130

Fig. 1 / *Sigillaria elegans* (Sternberg) (F-292). Harbour seam, Lingan Mine (p. 86); detail of favularian structure.

Fig. 2 / *Sigillaria elegans* (Sternberg) (F-349). Same location as above; specific identity of this specimen is uncertain but it is closely related to *S. elegans*.

1 (× 10)

2 (× 0.4)

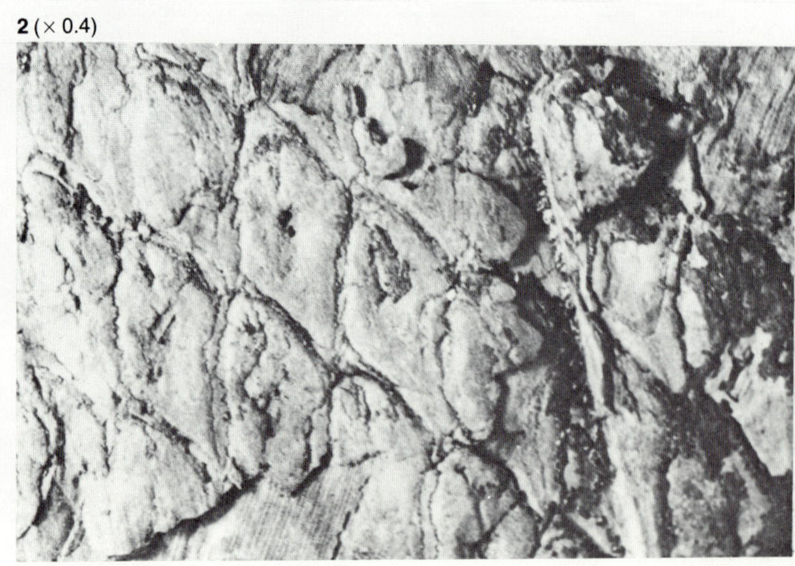

1 (× 2)

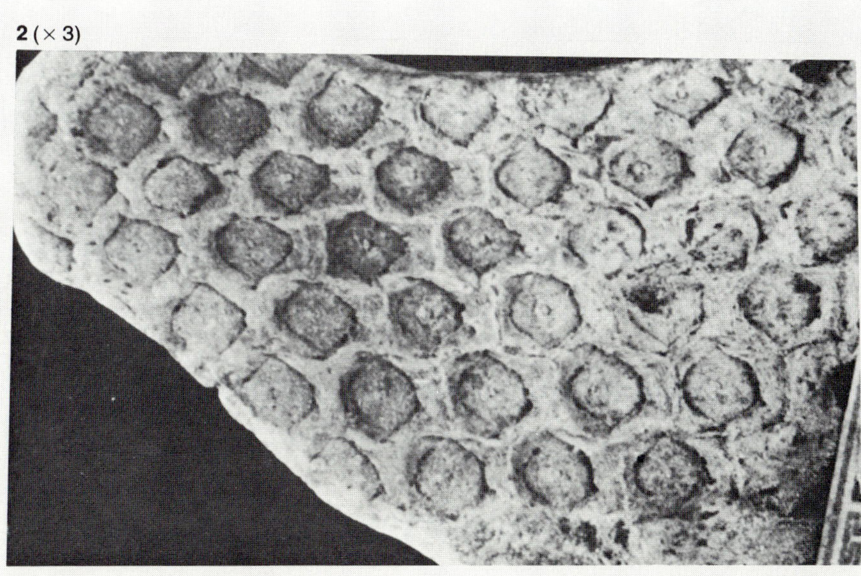

2 (× 3)

Plate 131

Fig. 1 / *Sigillaria elegans* (Sternberg) (F-433). Harbour seam, Lingan Mine, (p. 86); the specific identity is questionable.

Fig. 2 / *Sigillaria elegans* (Sternberg) (967G10.81). Morien series, Cape Breton Island; favularian structure is visible.

Plate 132

Fig. 1 / *Sigillaria laevigata* Brongniart (F-309). Harbour seam, Lingan Mine (p. 87); the mode of preservation of this specimen is known as the syringodendron condition (see Plates 135 and 136).

Fig. 2 / *Sigillaria laevigata* Brongniart (F-366). Same location as above.

1 (× 1.6)

2 (× 2)

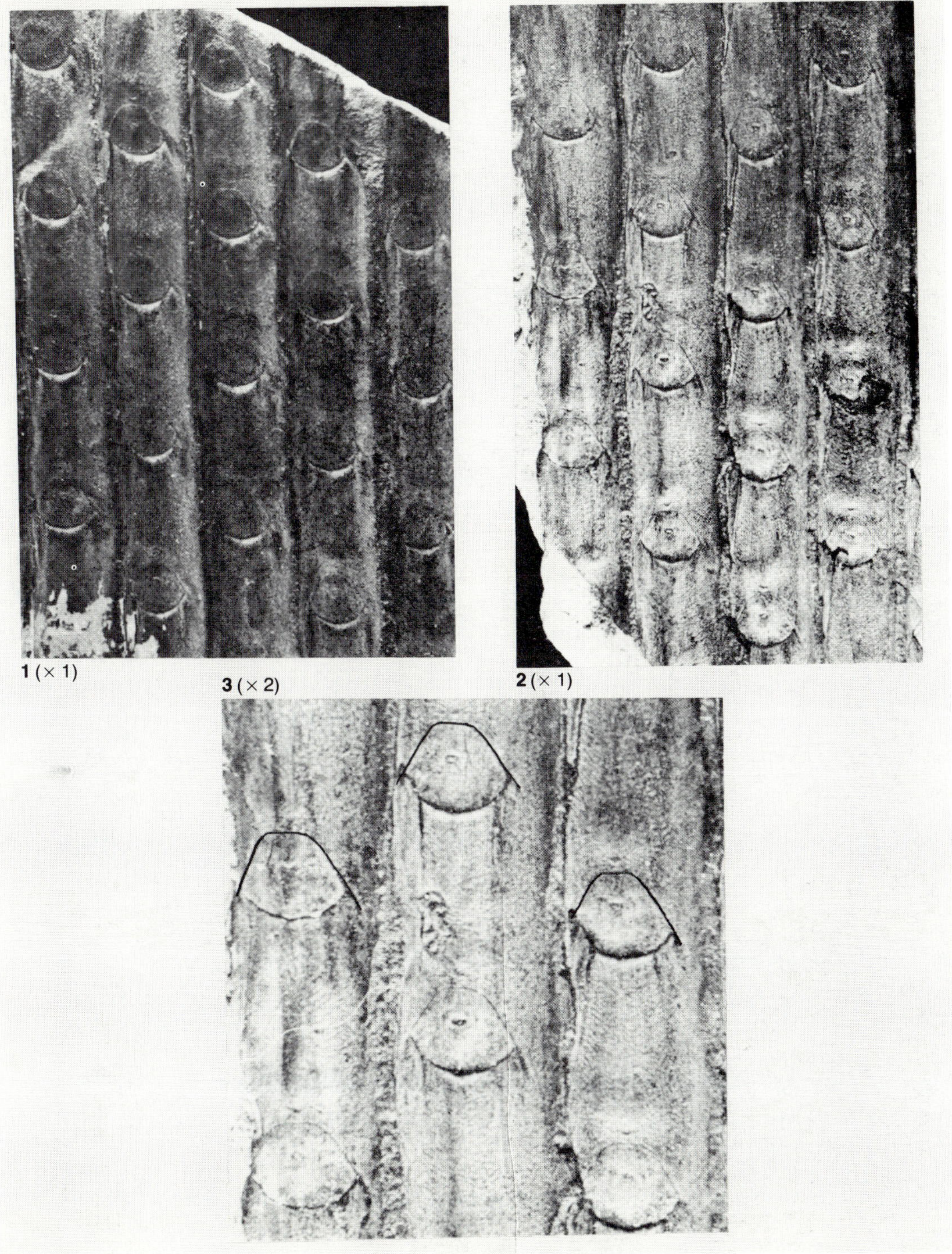

Plate 133

Fig. 1 / *Sigillaria scutellata* Brongniart (967G31.6). South Joggins, Cumberland Co., N.S. (p. 87).

Fig. 2 / *Sigillaria scutellata* Brongniart (967G31.6). South Joggins, Cumberland Co., N.S. Figs. 1 and 2 show different portions of the same fossil.

Fig. 3 / Detail of (967G31.6) showing characteristic leaf scars.

Plate 134

Fig. 1 / *Sigillaria* sp. (F-437). Harbour seam, Lingan Mine (p. 88). This is possibly a new species.

Fig. 2 / Detail of (F-437) showing faviform leaf cushions without separating ridges; infrared reflection image.

1 (× 2)

2 (× 3)

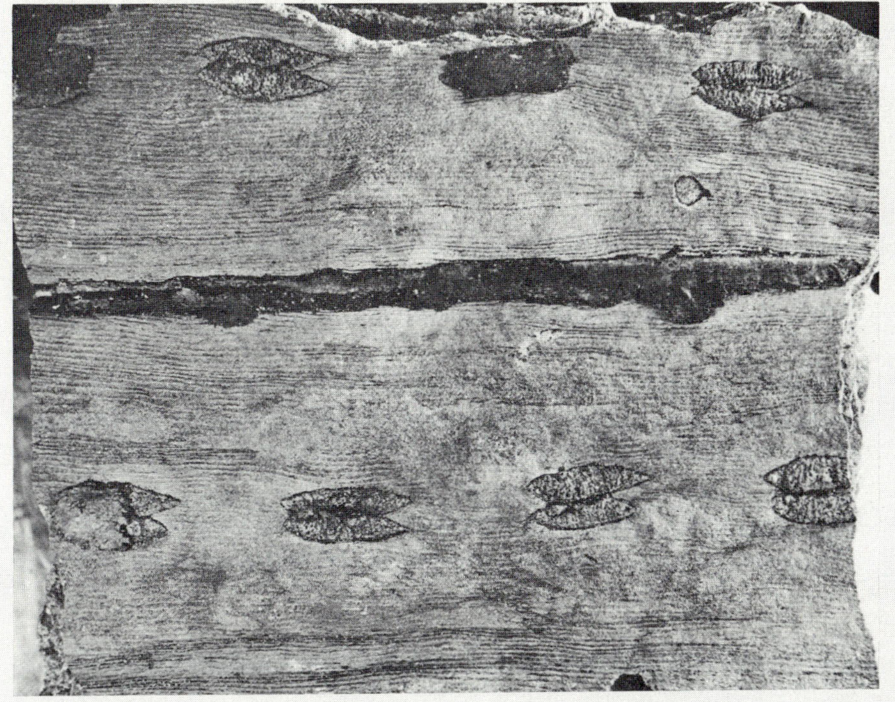

1 (× 1)

2 (× 5)

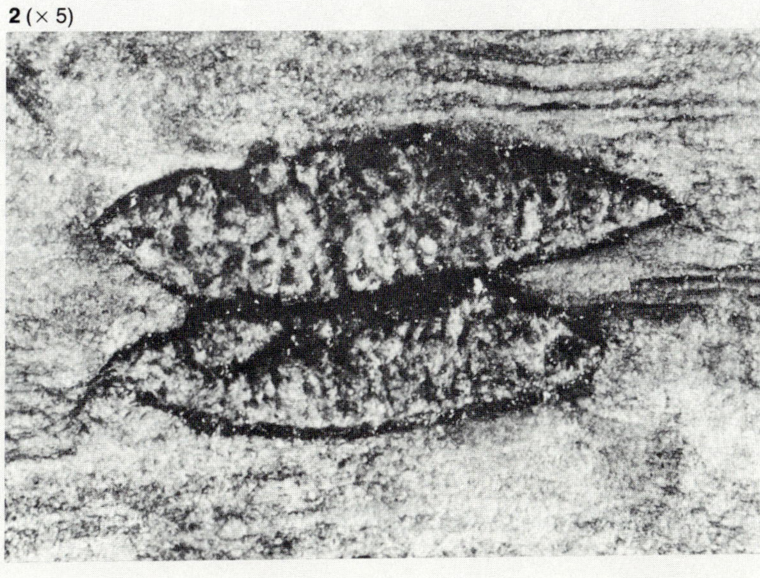

Plate 135

Fig. 1 / Sigillaria sp. indet. (F-352). Harbour seam, Lingan Mine (p. 88); the entire specimen is in a state of syringodendron preservation. (see Plate 132).

Fig. 2 / Detail of (F-352) showing a pair of parichnos scars under infrared reflection; see also Plate 136.

Plate 136

Fig. 1 / Detail of *Sigillaria* sp. indet. (F-352) showing a pair of parichnos scars partially covered with coalified material representing the tough outer layer of the tree (Gothan and Weyland, 1973, p. 156). (See Plate 135, Fig. 1, for the entire specimen.)

Fig. 2 / *Sigillaria* sp. indet. (F-313). Harbour seam, Lingan Mine (p. 88).

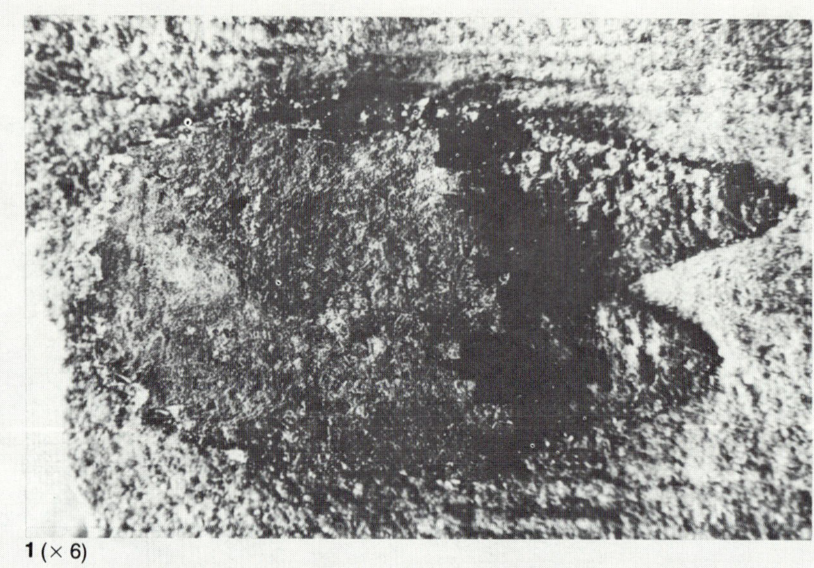

1 (× 6)

2 (× 1)

1 (× 1)

2 (× 0.5)

Plate 137

Fig. 1 / *Sigillaria tessellata* (Steinhauer) (F-459-3). Emery seam, Glace Bay (p. 88).

Fig. 2 / *Sigillaria tessellata* (Steinhauer) (861G1.1). Sydney Mines.

Plate 138

Fig. 1 / *Sigillaria tessellata* (Steinhauer) (967G10.90). Morien series, Cape Breton Island (p. 88).

Fig. 2 / *Lepidostrobus* sp. indet. (F-563). Emery seam, Glace Bay (p. 85); a complete cone in which the center is partially defoliated. (See next plate for additional detail of the cone.)

Fig. 3 / Detail of (F-563) showing the crown section of the strobus by infrared reflection. The sporophylls (left side of the photograph) are well preserved and acicular.

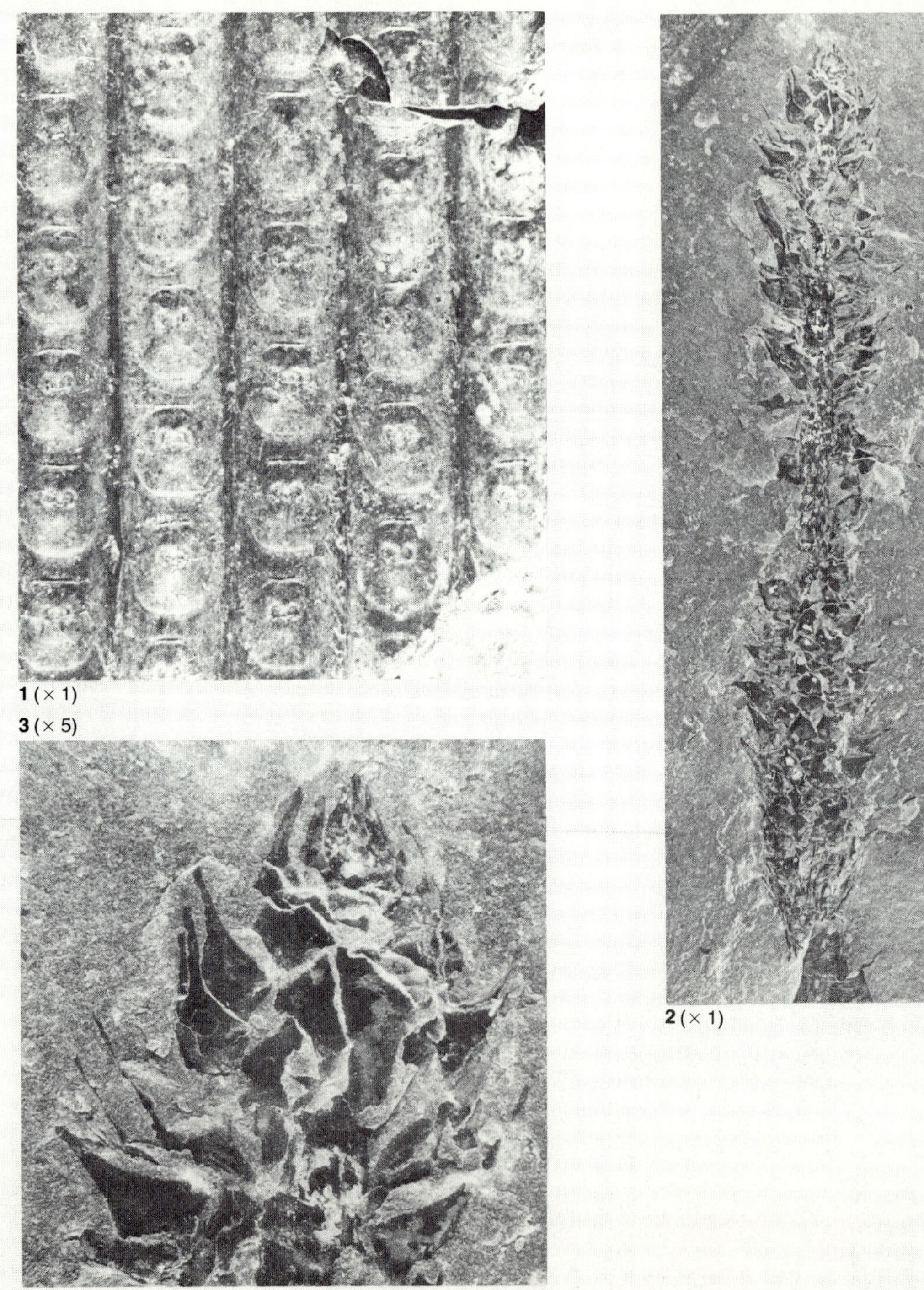

1 (× 1)

3 (× 5)

2 (× 1)

1 (× 7)

2 (× 1.3)

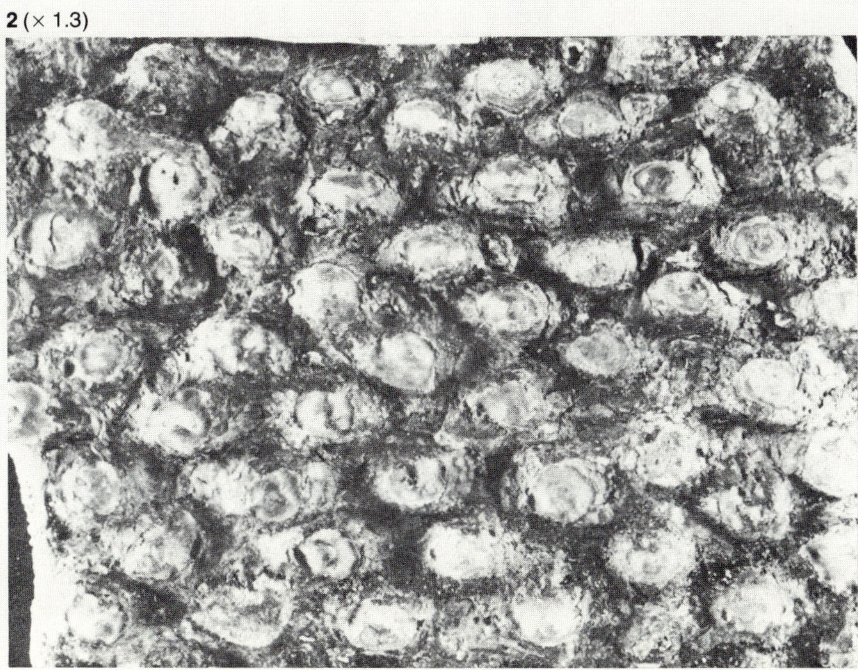

Plate 139

Fig. 1 / *Lepidostrobus* sp. indet., (F-563). Detail of the defoliated center part of the strobus showing leaf scars which are roughly vertical in arrangement. Note the broad base of the sporophylls (center part of the photograph). Infrared reflection image. (See Plate 138, Fig. 2 for the entire specimen.)

Fig. 2 / *Stigmaria ficoides* (Sternberg) (967G12.1). Drummond Colliery, Westville, Pictou Co., N.S. (p. 89).

Plate 140

Fig. 1 / *Stigmaria ficoides* (Sternberg) (F-351). Lingan Mine (p. 89). Arrow on the photograph indicates dichotomizing area of the rhizophore, which has been broken off and lost.

Fig. 2 / *Stigmaria ficoides* (Sternberg) (F-606). Morien series, Cape Breton Island; vascular core of the main roots.

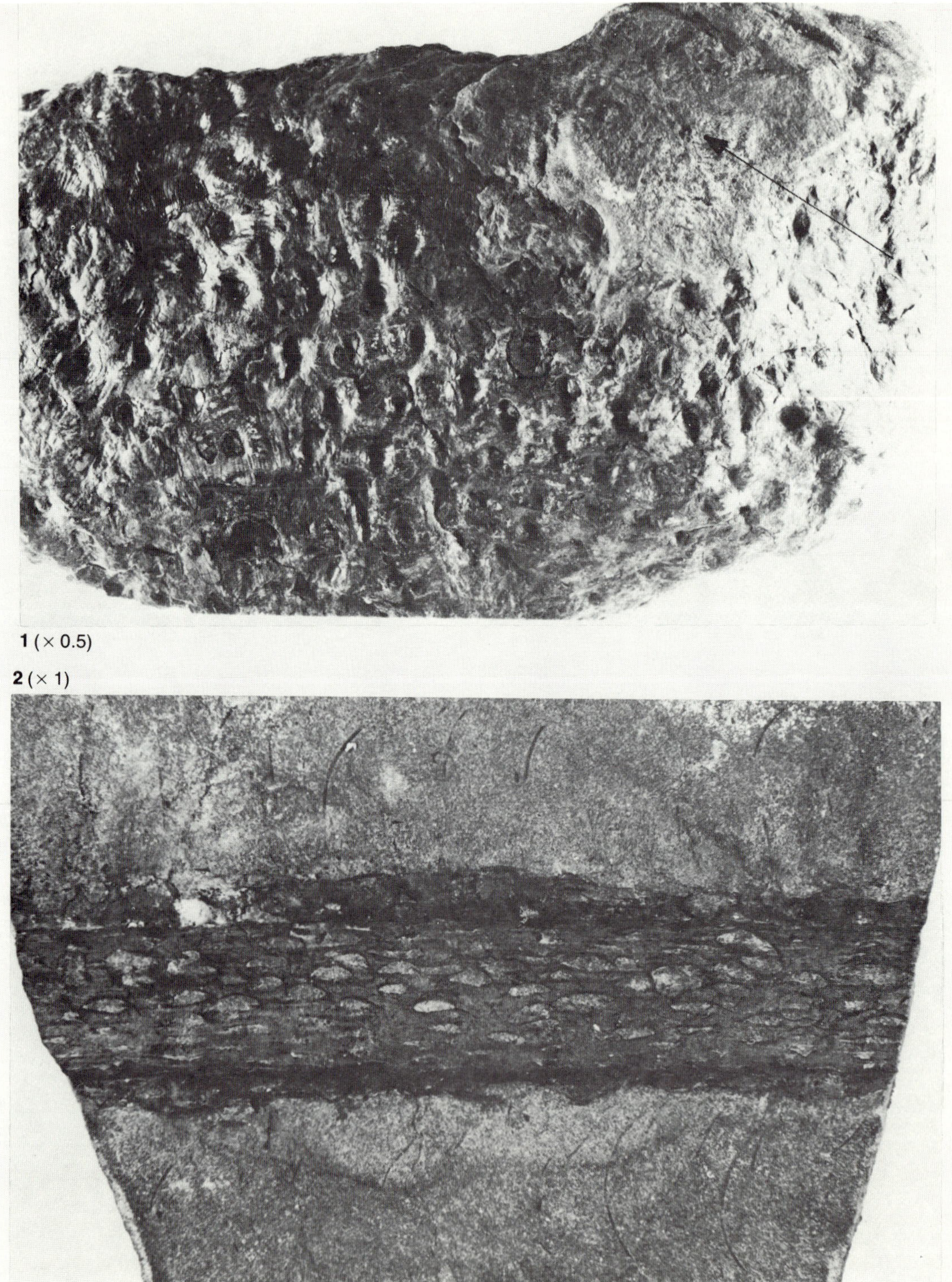

1 (× 0.5)

2 (× 1)

Plate 141

Fig. 1 / *Stigmaria* sp. indet. (F-358). Harbour seam, Lingan Mine (p. 90); dichotomizing system of stigmarians close to the trunk.

Plate 142

Fig. 1 / *Stigmaria* sp. indet. (F-357). Harbour seam, Lingan Mine, (p. 90); stigmarian appendages (true roots extending into sediments).

Fig. 2 / *Stigmaria* sp. indet. (F-353). Harbour seam, Lingan Mine; an uncommon type of stigmarian.

1 (× 0.66)

2 (× 2)

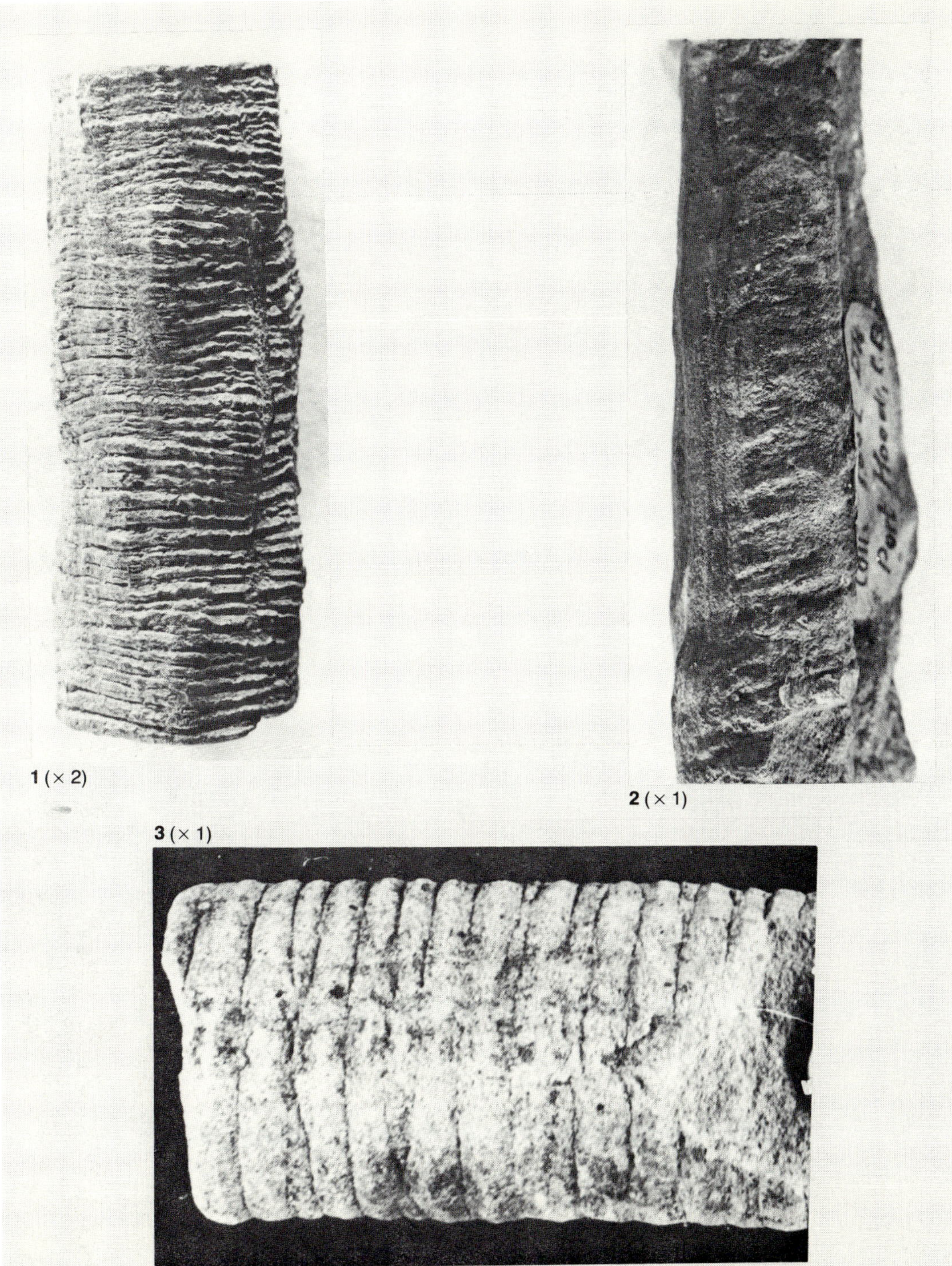

1 (× 2)

2 (× 1)

3 (× 1)

Plate 143

Fig. 1 / *Artisia* sp. indet. (976GF80.1) assumed to be from Nova Scotia (p. 90).

Fig. 2 / *Artisia* sp. indet. (967G123.13). Port Hood, Cape Breton Island.

Fig. 3 / *Artisia transversa* (Artis) (967G11.1). Collected by Dr. H. S. Poole on the shore of McCarren's Brook, South Joggins, Cumberland Co., N.S. (p. 90).

Plate 144

Fig. 1 / *Sigillariophyllum* sp. (F-623-1). Phalen seam, #26 Colliery, 1-B Mine (p. 89); strong median veins are well displayed in the leaves.

1 (× 1.5)

Plate 145

Fig. 1 / *Cordaites principalis* (Germar) (967G20.6), Stellarton series, McLellan Brook, Pictou Co. (p. 91).

Fig. 2 / *Cordaites* sp. indet. (F-355), Harbour seam, Lingan Mine (p. 91).

Plate 146

Fig. 1 / *Cordaites* sp. indet. (F-65). Mc Aulay seam (p. 91).

Fig. 2 / Megaspores (F-120-2) of uncertain affinity from the Emery seam in Glace Bay (p. 92); on block with F-120.

Fig. 3 / *Triletes* cf. *auritus* var. *grandis* Zerndt (F-120-1). Emery seam, Glace Bay. See Plate 54, Fig. 1, for location of this species on block F-120, (p. 92).

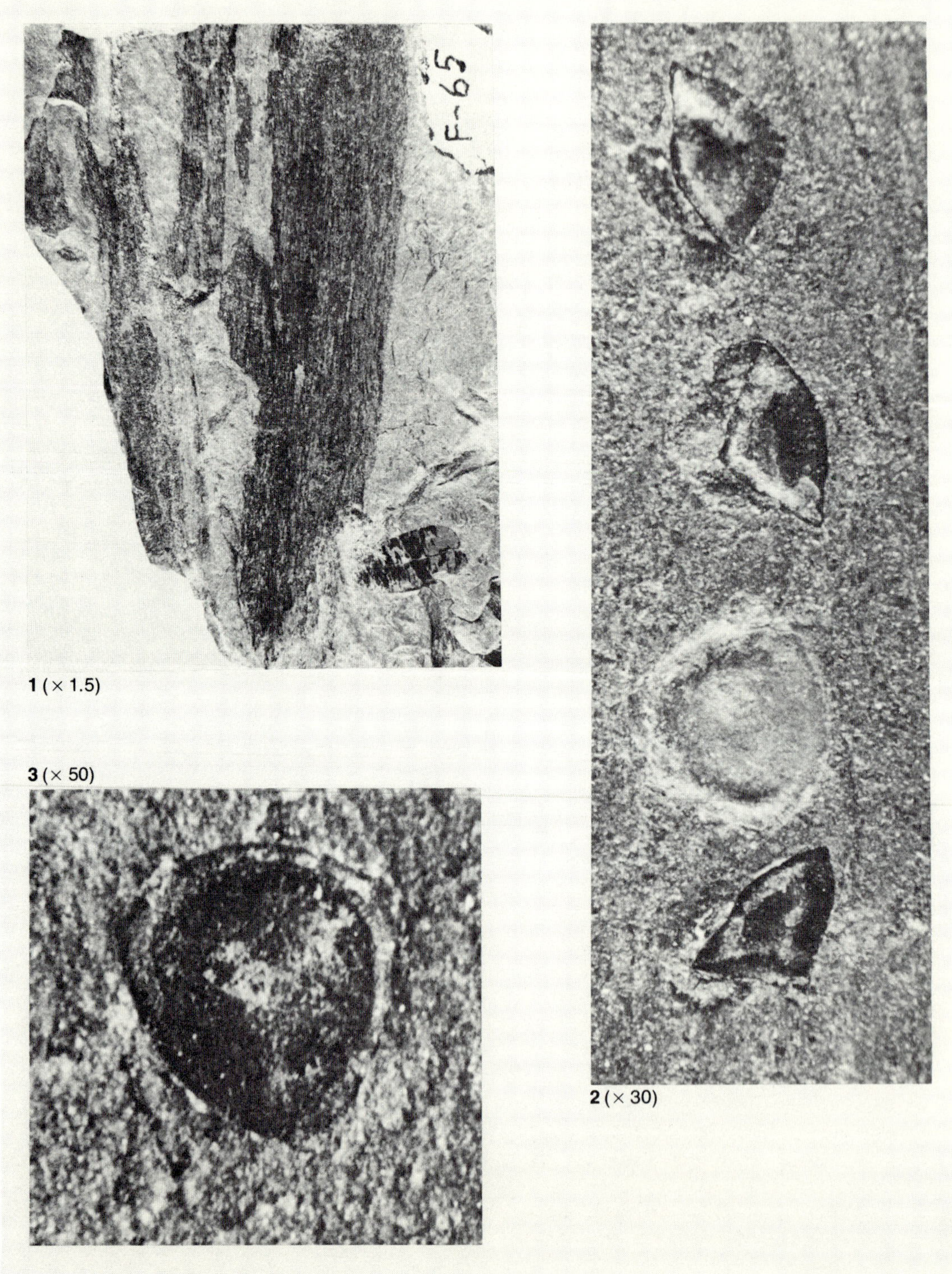

1 (× 1.5)

3 (× 50)

2 (× 30)

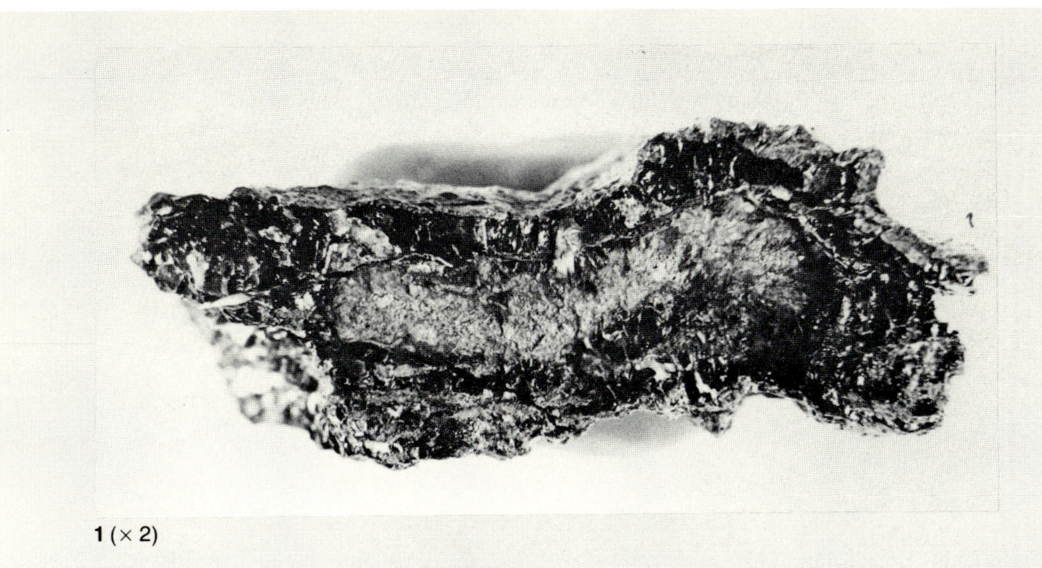

1 (× 2)

2 (× 6)

Plate 147

Fig. 1 / Unidentifiable plant fossil (F-83) with pyrite replacement in the core. Shoemaker seam. (Appendix I, p. 104).

Fig. 2 / Assorted coal samples from the Phalen seam, Glace Bay. The coal (dark areas) is dissected by veinlets of pyrite (light areas); infrared reflection image. (Appendix I, p. 105).

Plate 148

Fig. 1 / Stigmarian replaced by pyrite (F-244). Stubbart seam, Prince Mine, Point Aconi (Appendix I, p. 105).

Fig. 2 / Pyrite from the Mc Aulay coal measure (infrared reflection image) (Appendix I, p. 104).

1 (× 2)

2 (× 3)

1 (× 3)

Plate 149

Fig. 1 / Coal with oxidized pyrite (white areas) (F-382) from the Harbour seam of Lingan Mine (Appendix I, p. 105).

Plate 150

Fig. 1 / Plant matter (coal) now completely replaced by pyrite (F-406) (whitish material in the photograph); the specimen came from the roof of the Bonar seam (Appendix I, p. 105).

1 (× 3)

INDEX TO BOTANICAL SPECIES

Page number in bold type signifies primary entry of collected species; included is the St. Francis Xavier University Collection in Appendix II.

*Means that this specimen belongs to the Collection of the Department of Geology, St. Francis Xavier University, Antigonish, N.S., Canada; see Appendix II.

▶ Not previously reported from Sydney Coalfield (p. 10). Compare with Bell's records (1962a). This applies only to the specimens in the collection of the Nova Scotia Museum.

1 *S. hoeninghausi* is cited as *Lyginopteris (Sphenopteris) hoeninghausi* and similarly, *Lyginopteris oldhamia* as *Lyginopteris (Lyginodendron) oldhamia*, by Gothan and Weyland (1973, p. 295).

Addendum to INDEX OF BOTANICAL SPECIES: late entries not catalogued in the main section

▶ *Aphlebia* sp. (F-521, and on block F-592). Both from the Emery seam. Classifiable by Table 1; see Gothan and Weyland (1973, p. 230).

Boweria cf. *schatzlarensis* Kidston (F-634). Harbour seam, Lingan Mine. Classifiable by Table 2 (a fern); affinity unknown.

Dactylotheca plumosa (Artis) Brongniart forma *dentata* Brongniart (= *Senftenbergia dentata* (Brongniart)) (F-520). Emery seam. Classifiable by Table 2 (marattiaceous fern?).

Hymenotheca dathei H. Potonié (F-645). Stubbart seam, Prince Mine. Classifiable by Table 2 (a fern).

Pecopteris obtusa Bell (Bell, 1938, p. 80-81). (F-499, F-502): Harbour seam, Sydney Mines. (F-514): Emery seam. May be classifiable by Table 2 (a fern); should be classified by Table 1 (pteridophyll).

▶ *Sigillaria* sp. (F-625). 26 Colliery, 1-B Mine, Phalen seam. Certainly newly found in Cape Breton; may be an undescribed new species. See also F-437 in the catalogue section.

Lepidophloios sp. indet. (F-650). Stubbart seam, Prince Mine.

▶ *Sigillaria* sp. (F-638). Harbour seam, Lingan Mine. No *Sigillaria* sp. like this is figured by Bell (1938) or Crookall (1966, v. 4, Pt. 4).

Sphenopteris sp. (F-646). Stubbart seam, Prince Mine. Specific identification not possible at this time (1977).

INDEX TO MINERALS

Corrigenda

For *Asterophyllites equisetiformis* (Sternberg) Brongniart
read *Asterophyllites equisetiformis* (Schlotheim) Brongniart

For *Neuropteris (Mixoneura) obliqua* (Brongniart) Zeiller
read *Neuropteris (Mixoneura) obliqua* (Brongniart) Goeppert

For forma *angustifolia*
read forma *angustifolia* White

For *Neuropteris scheuchzeri* Hoffmann var. cf. *dawsoni*
read *Neuropteris scheuchzeri* Hoffmann var. cf. *dawsoni* Hartt

For *Pecopteris plumosa* Artis
read *Pecopteris plumosa* (Artis) Brongniart